AF618048

BAYESIAN METHODS IN RELIABILITY

TOPICS IN SAFETY, RELIABILITY AND QUALITY

VOLUME 1

Aims and Scope. Fundamental questions which are being asked these days of all products, processes and services with ever increasing frequency are:–

How safe?
How reliable?
How good is the quality?

In practice none of the three topics can be considered in isolation as they often interact in subtle and complex fashions. The major objective of the series is to cover the spectrum of disciplines required to deal with safety, reliability and quality. The texts will be of a level generally suitable for final year, M.Sc and Ph.D students, researchers in the above fields, practitioners, engineers, consultants and others concerned with safety, reliability and quality.

In addition to fundamental texts, authoritative "state of the art" texts on topics of current interest will be specifically commissioned for inclusion in the series.

The special emphasis which will be placed on all texts will be, readability, clarity, relevance and applicability.

The titles published in this series are listed at the end of this volume.

Bayesian Methods in Reliability

edited by

P. SANDER and R. BADOUX
TU Eindhoven, The Netherlands

SPRINGER SCIENCE+BUSINESS MEDIA, B.V.

Library of Congress Cataloging-in-Publication Data

Bayesian methods in reliability / edited by P. Sander and R. Badoux.
p. cm. -- (Topics in safety, reliability, and quality ; v. 1)
Includes bibliographical references and indexes.
ISBN 978-0-7923-1414-1 ISBN 978-94-011-3482-8 (eBook)
DOI 10.1007/978-94-011-3482-8
1. Reliability (Engineering) 2. Bayesian statistical decision theory. I. Sander, P. II. Badoux, R. III. Series.
TA169.B39 1991
620'.00452--dc20 91-31375

ISBN 978-0-7923-1414-1

Printed on acid-free paper

CONTENTS

INTRODUCTION

These proceedings contain the core material of a course that every two years is organised by the European Safety and Reliability Association, the Faculty of Industrial Engineering and Management Science of the Eindhoven University of Technology, and the Dutch Society for Reliability Technology. The course was run for the first time in Eindhoven (NL) in October 1988 and was repeated in Bradford (UK) in October 1990. The course is one of the modules that is accepted by the University of Bradford as partial fulfilment of the requirements for the Master Programme in Safety and Reliability. For more information about this programme we refer to Dr. A.Z. Keller, University of Bradford, Department of Industrial Technology.

The proceedings include six papers. The first paper, by Badoux is a general introduction in which it is explained that there are two reasons why Bayesian methods are essential in reliability. The first reason is sparse data, which is a consequence of highly reliable equipment, and the second reason is that Bayesian methods give a sound basis to the natural logic of the decision maker. The introduction is completed by two real life examples.

In the second paper Smith presents a detailed discussion of the Bayesian statistical methods. First, a review of some basic probability concepts is given. Then, some logical and practical difficulties with the non-Bayesian statistical approach are noted. This leads to a presentation of the mechanics of Bayes theorem, as a procedure for combining judgements and data in order to learn from experience. Finally, some recent progress towards the computational implementation of Bayesian methods is reviewed and illustrated.

In chapter 3 Newby presents some simple models for the analysis of censored data from non-repairable systems. Some examples illustrate the use of these models in reliability estimation. In the next chapter Newby proceeds with repairable systems and growth models. The renewal process, the homogeneous Poisson process and the non-homogeneous Poisson process are discussed. Also the Duane model is presented. The value of Bayesian and graphical methods in analysing data from repairable systems is illustrated.

In chapter 5 the use of expert judgement in risk assessment is explained. The problem is, again, that sometimes sufficient hard data may be unavailable. In such cases risk assessment can only be based upon the judgement of experts who draw on their knowledge and experience of failures in related but substantive different areas. In this chapter some of the mathematical and statistical issues which arise in eliciting and combining expert judgements of the likelihood of particular events or uncertain quantities are indicated.

In the last chapter, Littlewood presents the latest view on the problem of forecasting software reliability. The software failure process is a non-stationary stochastic process. Several of the best-known software reliability growth models are described, and their performance on real software failure data are compared. The conclusion is that the predictive quality of a model must be tested comparing past predictions emanating from the model with the actual behaviour for a particular data set. Only when this comparison shows nothing but minor differences, one can have confidence in future predictions for the same data.

The Editors

1. INTRODUCTION TO BAYESIAN METHODS IN RELIABILITY

by

ROBERT A.J. BADOUX[1]

Abstract

The reasons for applying Bayesian methods in reliability problems are given. Bayes' theorem is explained and the Bayes' technique is demonstrated by means of an example. Two further examples taken from a safety study on gas transmission pipelines show the versatility of Bayesian methods.

1. Why Bayesian Methods?

There are several reasons for applying Bayesian methods in reliability problems. However there are two main reasons. The first one is sparse data. The second reason is somewhat more complicated and stems from decision theory. Both reasons are explained in more detail.

1.1. Sparse data

In most reliability problems one has to make inferences on only a few data. These inferences mostly concern parameters of failure distributions and quantities such as availability, reliability and probability of failure on demand.

Because most equipment is highly reliable one encounters not many failures in practice. Even with a large amount of equipment in use one may be unable to make inferences. The reason for this is the heterogeneity of the data. The analysis of field data often shows that equipment is used under different circumstances (temperature, humidity, pressure, load etc.). Also different maintenance schemes for the same type of equipment prevents the pooling of data.

---

[1] presently: N.V. Nederlandse Gasunie

P. Sander and R. Badoux (eds.), Bayesian Methods in Reliability, 1–13.

Life testing is rather expensive and time consuming. Therefore one performs accelerated life tests or one makes use of censoring. Design changes in the equipment for instance may make it impossible to use results of earlier tests. Classical statistical methods give no solutions to these problems. The answer is found in applying Bayesian techniques. These make it possible to pool all kinds of information in a uniform and consistent way. This also applies to sound engineering judgement being taken into account.

The principle of Bayes' theorem, also known as Bayes' rule, is depicted in Figure 1. Consider the problem of updating information on the hazard rate λ (cf. Smith section 4) of a certain type of equipment. Interpretation of data of similar equipment on engineering grounds results in a prior distribution. The combination of this prior information with field data through Bayes' rule results in a posterior distribution giving new information about λ (cf. Smith sections 11 and 12).

In the updating process this posterior distribution will serve next time as prior information. The process can be seen as an iterative process. The updating process converges theoretically to perfect information about λ. By that time this type of equipment probably will be technically spoken out of date and replaced by a completely new technical concept. Bayes' theorem is explained in section 2. (A more complete discussion of the Bayes principle can be found in chapter 2).

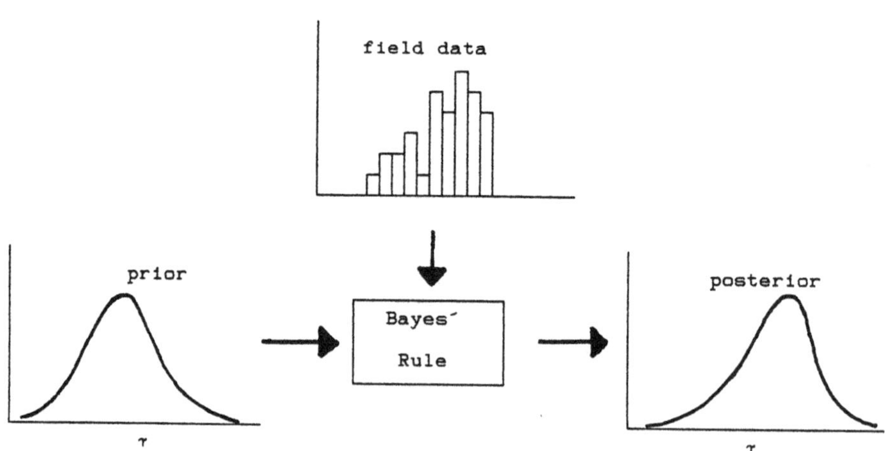

Figure 1. The principle of Bayes' Theorem.

1.2 Decision problems

A reliability analysis is always performed to give an answer to some kind of decision problem. The consequences of the decisions based on estimates of parameters often involve money or, more generally, some form of utility. Hence the decision maker is more interested in the practical consequence of his estimate than in its theoretical properties. In particular, he may be interested in making estimates that minimize expected loss (cost). This is demonstrated by means of an example.

Consider the following problem. A designer can choose between two system configurations. The configurations being:

A. two boiler–feedwater pumps, one of 100% capacity and one of 50% capacity, and

B. three boiler–feedwater pumps, each of 50% capacity.

The two systems are depicted in Figure 2.

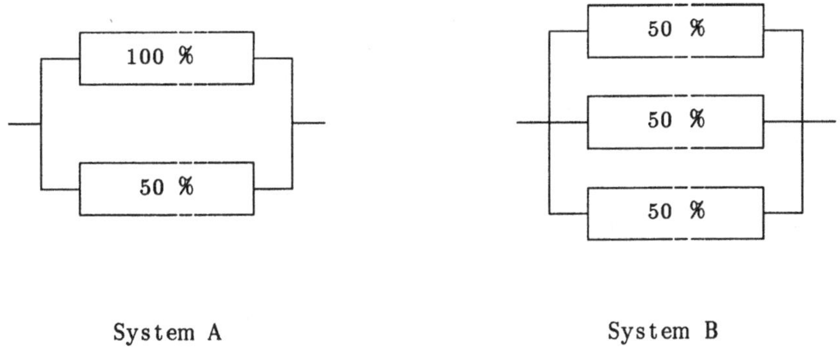

Figure 2. Two system configurations.

Configurations	Full Load	Half Load
System A	1 v 1	1 v 2
System B	2 v 3	1 v 3

Table 1. Types of system at different loads.

The load pattern for both systems is the same, viz. full load for 60% of the time (on average) and half load for 40% of the time (on average). This means that the system configurations differ with the load pattern. The different situations are given in Table 1. (Notation: k v n means k–out–of–n).

From Table 1 it can be deduced that failure of one or more pumps will result in partial or full loss of power. So, the loss function of a system is a function of the unavailability of the pumps and the corresponding loss of power. The expression for the unavailability of a pump contains parameters such as the hazard rate λ. In a Bayesian context this parameter is a random variable. In that way we can take any uncertainty in λ into account in the decision to be taken. So the loss function becomes a random variable.

Now let L(A) be the loss function for system A and L(B) is the loss function of system B. The investment for system B is higher than the investment for system A. This has been accounted for in the loss function of both systems. The decision between two system configurations is based on the difference

$$L(A-B) = L(A) - L(B)$$

If

$$P[L(A-B) \leq 0] > 50\%$$

then system A is to be preferred, else system B is the better option. The situation described above is depicted in Figure 3.

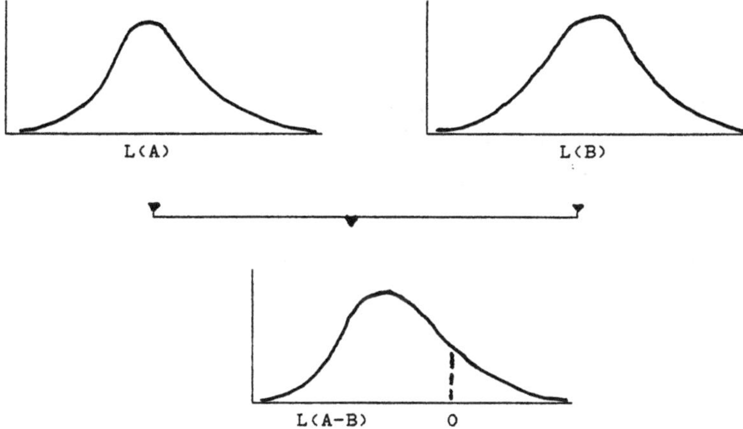

Figure 3. The distribution of L(A–B).

2. Bayes' Theorem

Bayes' theorem provides a means to adjust the probability of occurrence of an event to incorporate additional data. A basic axiom of probability theory (which reflects the meaning of probability), states that the probability of two simultaneous propositions, A and B, is given by

$$P(A \cap B) = P(A) \cdot P(B|A) \tag{1}$$

and by

$$P(A \cap B) = P(B) \cdot P(A|B) \tag{2}$$

Equating the right side of both equations and dividing by P(B) gives

$$P(A|B) = P(A) \left[\frac{P(B|A)}{P(B)} \right] \tag{3}$$

which says that P(A|B), the probability of A given B, is equal to P(A), the probability of A prior to having information B, times the correction factor given in the bracket.

The operation of Bayes' rule is shown by a numerical example.

Assume that the plot in Figure 4 represents the distribution of the mean time between failure, MTBF, θ based on expert opinion. The curve actually represents prior knowledge (cf. Smith section 11).

Suppose the new data, event B, to be 5 failures in 375 months, then the MTBF-analysis becomes

$$P(\theta_i|B) = P(\theta_i) \frac{P(B|\theta_i)}{P(B)} \tag{4}$$

$P(\theta_i|B)$ = Probability of MTBF θ_i, given information B

$P(\theta_i)$ = Probability of MTBF θ_i, prior to having information B

$P(B)$ = Probability of B

$P(B|\theta_i)$ = Probability of B, given that the MTBF is θ_i

θ_i = MTBF for a discrete time interval.

The equation for $P(B|\theta_i)$ is as follows:

$$P(B|\theta_i) = \frac{(T/\theta_i)^n}{n!} \; exp \; \{-T/\theta_i\} \tag{5}$$

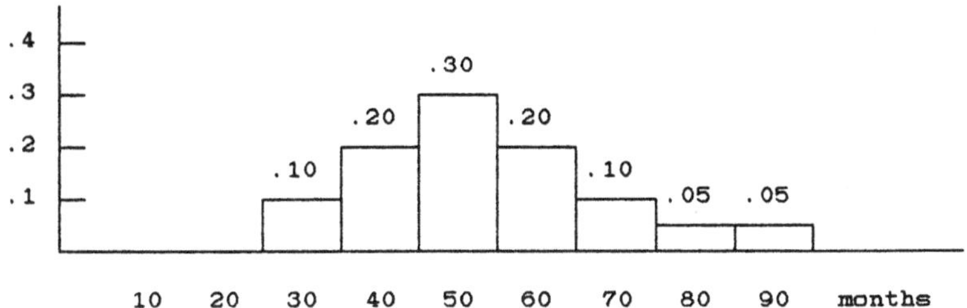

Figure 4. The prior distribution of the MTBF θ.

Here is:

n = Number of events (failures)

T = Number of time periods

θ = MTBF

Using the new data ($n = 5$ failures in time period $T = 375$ months) the equation for this example becomes:

$$P(B|\theta_i) = \frac{(375/\theta_i)^5}{5} \, exp\{-375/\theta_i\} \qquad (6)$$

θ_i	30	40	50	60	70	80	90	Σ
$P(\theta_i)$	.1	.2	.3	.2	.1	.05	.05	1.0
$P(B\|\theta_i)$	.00948	.05119	.10937	.15342	.17334	.17369	.16226	–
$P(\theta_i)P(B\|\theta_i)$	.00095	.00335	.03281	.03068	.01733	.00868	.00811	.1019
$P(\theta_i\|B)$	.00932	.03281	.32194	.30104	.17005	.08517	.07958	1.0

Table 2. MTBF probability.

The prior data given in Figure 4 are tabulated as shown in Table 2. The MTBF values listed are placed in row 1. The probability values for those MTBF values are placed in row 2. The values of $P(B|\theta_i)$ are calculated using equation (6) for all the values of MTBF in the top row and then placed in row 3. The values of row 4 are calculated by multiplying the values in row 2 with the values in row 3. The values in row 4 represent the posterior value except they must be apportioned to have the sum of the probabilities equal to 1. This is done by summing the values in row 4 and recording the sum in the right column. The individual values in row 4 are then divided by the sum of row 4 with the results recorded in row 5.

Row 5 of Table 2 is the final probability distribution or posterior distribution after having the information B combined with the prior distribution. The results are displayed graphically in Figure 5. The evidence indicates the MTBF probability distribution will shift slightly to higher values when the new data are considered.

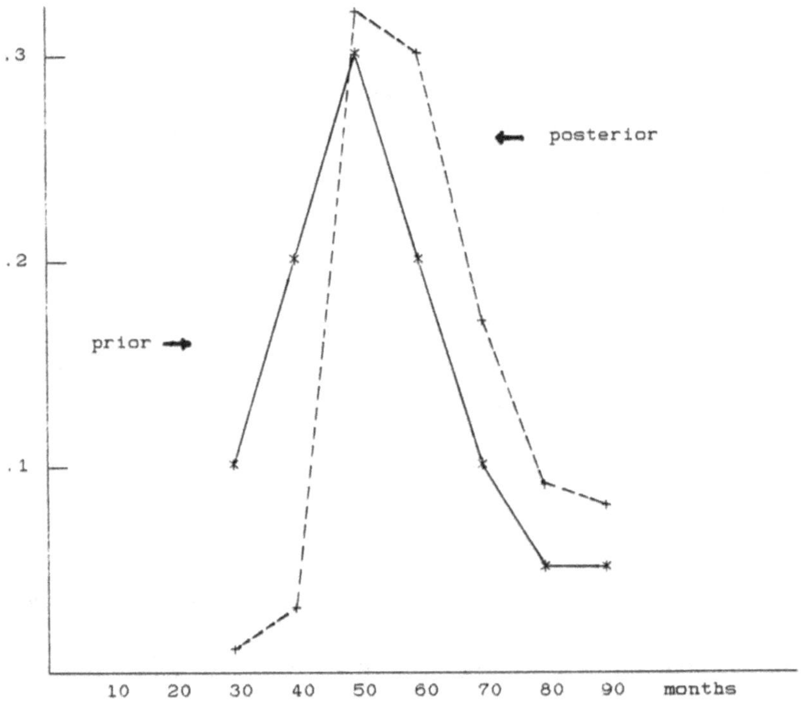

Figure 5. Prior and posterior distribution of the MTBF θ.

3. Examples from a Safety Study on Gas Transmission Pipelines

In the following two examples the practical application of Bayesian techniques in estimating probabilities is shown. Both examples are about pooling of historical accident data coming from different sources. This asks for a Bayesian approach. The reference for the data in these examples is a safety assessment study carried out by TNO on behalf of the Dutch Ministry of Public Health and Environmental Protection (see TNO, 1982).

3.1 Estimating the probability of the development of a big hole

The first example concerns the estimation of the probability of the development of a big hole in a gas transmission pipeline given a leak caused by an external force. The gas transmission pipeline is part of the Dutch regional distribution grid.

Dutch data show that over the period 1975–1980 there were 8 leaks caused by external forces. Of these 8 leaks the size of the hole was known in only 5 cases. On two occasions the hole was to be considered as big. This means that for a first estimate for the probability of a big hole one may use $p_1 = .40$ on the basis of the Dutch data.

U.S. data over the period 1970–1978 show 412 leaks caused by external forces. In 48 cases there was a big hole. A second estimate for the probability of a big hole comes down to $p_2 = .1165$ on the basis of the U.S. data. Now there is a problem because p_1 and p_2 differ significantly.

The data on the 412 leaks caused by external force are based on the so-called '20-day reports' with a total of about 2000 leaks over the period 1970–1978. The 'annual reports' show a number of about 200,000 leaks over the same period. This implies that the number of leaks caused by an external force with a small hole is probably much higher than may be concluded from the data on the basis of the 20-day reports. The U.S. data are quite comprehensive, however the completeness and the reliability of the data are rather doubtful.

The estimate for the probability of a big hole is for use in the Dutch situation. For this reason the U.S. data are considered as prior information. The Dutch data are actual field data although of limited size.

Let θ be the probability of a big hole given a leak caused by external force. This θ is the parameter of a binomial probability model. The natural prior for θ in this case is a Beta distribution (cf. Newby section 3.2 and Mann *et al.* (1974)). Let the parameters of this distribution be α and β.

The mean value of α is:

$$E(\theta) = \frac{\alpha}{\alpha + \beta} \tag{7}$$

The U.S. data is appraised in the following sense:

- as a conservative approximation of $E(\theta)$ is chosen

$$\frac{\alpha}{\alpha + \beta} = 1/15 = .0667$$

  which implies that $\beta = 14\,\alpha$.
- as 90%-point of the Beta distribution is chosen $p_2 = .1165$, i.e. $P(\theta \leq .1165) = .90$.

The above information results in the following values for α and β (see Mann *et al., loc. cit.*, page 394): $\alpha = 3$ and $\beta = 42$.

The posterior distribution under the binomial probability model is again a Beta distribution when the prior distribution is a Beta distribution (cf. Smith section 15). Applying Bayes' rule to the example leads to the following Bayesian estimate for the probability of a big hole given a leak caused by an external force:

$$E(\theta|k{=}2,\ n{=}5) = \frac{\alpha + k}{\alpha + k + n} = \frac{3 + 2}{3 + 42 + 5} = .10 \tag{8}$$

(k = number of leaks with a big hole, n = total number of leaks caused by an external force with known size of the hole).

The Bayesian two-sided 90% - credible interval is defined by:

$$P(.0410 \leq \theta \leq .1782) = .90 \tag{9}$$

(cf. Smith section 13 and Mann *et al.* (*loc. cit.*)).

3.2 Estimating the leak rate of a gas transmission pipeline

The second example concerns the estimation of the leak rate of a gas transmission pipeline belonging to the Dutch main distribution grid.

The Dutch data show only 1 leak over the period 1970–1975. The total experience over this period equals 23000 km*years. A first estimate for the leak rate would be:

λ_1 = 1/23000 km*year = .04/1000 km*year.

However, this estimate is based on only 1 leak over a 5 year period.

Analysis of U.S. data on leaks in comparable pipelines (i.e. coated and equipped with cathodic protection) results in about 1531 leaks over the period 1970–1978. The total experience in the U.S. over that period amounts to 3,100,000 km*years. A second estimate based on the U.S. data would be:

λ_2= 1,531/3,100,000 km*year = .49/1000 km*year.

The problem is that λ_1 and λ_2 differ greatly. The quality of the U.S. data is rather doubtful as has been demonstrated before. Therefore the only solution is to apply Bayes' rule after having appraised the U.S. data in a proper way.

Let λ be the leak rate. In this case λ is the parameter of the Poisson probability model. The usual prior distribution in this case is the Gamma distribution with parameters α and β (cf. Smith section 15). The mean value of λ is:

$$E(\lambda) = \beta/\alpha \qquad \text{(see Mann } et\ al.\text{)} \qquad (10)$$

The interpretation of the U.S. data comes down to:
- as estimate for $E(\lambda)$ is chosen

$$\beta/\alpha = .49 \qquad \text{(the applicable unit is 1/1000 km*year)}$$

- as 90% – point of the Gamma distribution is chosen

$$\beta/\alpha + 1.3 * \beta/\alpha = 1.13 \qquad \text{(i.e. } P(\lambda \leq 1.13) = .90\text{)}$$

The above information leads to the following values for α and β (see Mann *et al.*, p. 400): $\alpha = 2.04$ and $\beta = 1$.

The posterior distribution under the Poisson model is again a Gamma distribution when the prior distribution is a Gamma distribution. Applying Bayes' rule to this example results in the following estimate for the leak rate in a gas transmission pipeline:

$$E(\lambda|n=1,T=23) = \frac{\beta + n}{\alpha + T} = \frac{1 + 1}{2.04 + 23} = .08/1000 \text{ km*year} \tag{11}$$

(n = the number of leaks in pipelines of the main distribution grid, T = total experience in km*years).
The Bayesian two-sided 90% – credible interval is defined by:

$$P(.01 \leq \lambda \leq .19) = .90$$

4. Conclusions

It is shown that Bayesian methods can be a powerful tool in reliability analysis. Pooling data from different sources in a consistent and uniform way is possible. In particular in the case of sparse data it is possible to incorporate expert opinion.

Of course there is always the problem of fitting a proper prior distribution. The calculation of a posterior distribution can be very cumbersome when the posterior is not of the same type as the prior. However, there are powerful numerical methods to solve this problem (cf. Smith section 18).

In conclusion it can be said that Bayesian methods in reliability are a way of quantifying engineering judgement. In spite of all nice statistical techniques it should be kept in mind that there is an engineering problem to be solved.

References

MANN, N.R., SCHAFER, R.E., & SINGPURWALLA, N.D. (1974), *Methods for Statistical Analysis of Reliability and Life Data*, Wiley, New York.

TNO (1982), *Veiligheidsstudie betreffende het transport per ondergrondse pijpleiding van aardgas en LPG in Nederland (in Dutch)*.

N.V. NEDERLANDSE GASUNIE
P.O.BOX 19
9700 MA GRONINGEN
THE NETHERLANDS

2. AN OVERVIEW OF THE BAYESIAN APPROACH

by

ADRIAN F.M. SMITH

Imperial College London

Abstract

A framework is provided for combining the background knowledge and judgement of the subject matter expert with experimental or on-line data. It is shown that the logic of decision making points to the Bayesian approach as the natural one to deal with the various issues. A detailed discussion of the Bayesian statistical methods is given. Some recent progress towards the computational implementation of Bayesian methods is received and illustrated.

1. Background

Quantitative assessment of uncertainties pervades the study of the functioning of systems (or subsystems, or unit components). Whether we are concerned with *Reliability*, *Availability*, *Maintainability* or *Repairability*, we are essentially studying time-related probability assessments of successful functioning.

But how are such probability assessments to be made? What is the underlying "uncertainty logic" that should be used? What are the "raw materials" that go into the making of such assessments? And how do such assessments feed into the "decision-making" process?

Clearly, there is a need for concepts and tools for handling uncertainty. And for handling it in such a way that there is a natural linkage with rational decision making. Moreover, such concepts and tools must embrace both of the kinds of "raw materials" familiar to those confronting uncertainties in complex systems: on the one hand, the background knowledge and judgements of the subject matter expert; on the other hand, experimental or on-line data in the form of counts or measurements of failures or failure-times. How should

P. Sander and R. Badoux (eds.), Bayesian Methods in Reliability, 15–79.

these two types of input be combined? And how should one take account of the fact that, both with expert judgements and with test or operating data, experience has typically been derived from many different sources and contexts, whose degree of relevance to the situation currently under study may be questionable.

The material presented in this chapter is aimed at providing a framework of ideas and procedures for tackling these issues. We begin by reviewing some basic probability concepts and providing some notation. Aspects of particular concern in reliability studies are then identified, as are common forms of data description arising from test or operational studies. We then briefly discuss the problem of statistical learning about unknown aspects of failure distributions and give a short review of some classical (non-Bayesian) statistical approaches to treating such problems. Logical and practical difficulties with these approaches are noted, as a lead-in to a detailed discussion of Bayesian statistical methods. We begin this with a brief introduction to the logic of decision-making, noting, in particular, how this points to the Bayesian approach as the natural one to deal with the various issues discussed above. The mechanics of Bayes theorem, as a procedure for combining judgements and data in order to "learn from experience", is then illustrated, in both intuitive and mathematical terms. Methods for summarizing inferences and predictions are discussed, together with approximation techniques. The problem of combining data from many different sources is introduced and related to so-called "Empirical Bayes" methods. Finally, some recent progress towards the computational implementation of Bayesian methods is reviewed and illustrated.

The aim is to provide the reader with an overview of the What? Why? and How? of Bayesian statistics, as a methodology for thinking about, representing and updating uncertainty in a logical fashion. Later chapters will focus on applications of these ideas to reliability problems.

2. Probability Concepts

In everyday usage, we are familiar with the interpretation of the term "probability" in two quite distinct senses. On the one hand, in contexts of repeatable experiments the word is often used as it were synonymous with "relative frequency", and thus closely linked with the idea of physical variability. On the other hand, in contexts of "one-off" eventualities, where uncertainty is present but repeatability is rather meaningless, the word is often used as if it were synonymous with personal judgements of appropriate betting odds, reflecting personal "degrees of belief" about uncertain outcomes.

Luckily, for our purposes we do not have to worry about this "schizophrenic" attitude to "probability". From a mathematical point of view, the same "rules of probability" apply, no matter what the interpretation. And from an interpretative point of view, it will be very useful to be able to combine "frequency-type" inputs (such as failure time data from repeated test runs) with "degree of belief" inputs (such as expert judgements about an as yet untested component). We shall therefore review basic probability ideas having in mind that they may find application in either frequency or degree of belief contexts, or both in combination.

We shall assume that the reader is familiar with the following concepts. The starting point in a statistical analysis is a concern with *uncertain events*, uncertainty (representing physical variability, or personal beliefs, or both) being represented *quantitatively* by *probability.* When the uncertain events are defined in terms of as yet unobserved quantities (*counts* or *measurements*) we refer to the latter as *random variables* or *random quantities* (called *discrete* in the case of counts, *continuous* in the case of measurements). A detailed description of how we assume uncertainty to be distributed over ranges of possible outcomes is called a *probability distribution.* This is represented by a *mass function* in the discrete case and a *density function* in the continuous case. Summaries of distributions are typically given in terms of *location* and *spread.* Such summaries include the *mean* and *standard deviation* (the square root of the *variance*) as well as *percentiles* of the distribution. When data has been collected and summarized (perhaps as a *bar chart*, *histogram* or *empirical cumulative distribution function*), we refer to an *empirical* distribution. When a mathematical model is specified to describe a mass or density function, we refer to a *theoretical* distribution.

Common forms of discrete, theoretical probability distributions include: the *Binomial* (counting the number of "successes" in a fixed number of independent success/failure trails); the *Geometric* (counting the number of trials until the first "success" in independent success/failure trials); the *Poisson* (counting the number of events occurring in a fixed interval of a "completely random" process). Common forms of continuous theoretical probability distributions include: the *Normal* (often arising as a model of a measurement, which is perceived to be the aggregate of a large number of independent perturbations); the *Exponential* (measuring the waiting time between occurrences in a "completely random" process).

When more than one random variable is being considered, the richness of the uncertainty description is considerably extended. The description of uncertainty about all the random variables *simultaneously* is called the *joint distribution.* The implied uncertainty descriptions for individual random variables are called (*univariate*) *marginal* distributions; the description of pairs of random variables (selected from the many) are called (*bivariate*) *marginal* distributions. The description of uncertainty for a subset of the random variables given specified values for some other subset of random variables is called a *conditional* distribution. If a joint probability description can be written as the product of marginal probability descriptions, the random variables are said to be *independent* (and vice versa). An analogous statement holds for *conditional independence.* When a single random variable is transformed, or a function is formed of several random variables, *transformation techniques* exist, whereby the probability description of the tranformed quantity can be deduced from the probability description of the original random variables.

3. Notation

We shall denote a random variable by $\tilde{x}$ and a realized (observed) value by x: thus, $\tilde{x} = x$ means that "the random variable $\tilde{x}$ turns out to have the value x".

The mass or density function for a random variable $\tilde{x}$ will be denoted by f(x); the cumulative distribution function by F(x), so that f = F', where F' denotes the first derivative of F. If $\tilde{x}_1, \ldots, \tilde{x}_n$ are n random variables, the joint density (or mass) function will be denoted by $f(x_1,\ldots,x_n)$. The marginal density for $\tilde{x}_i$, say, is then given by

$$f(x_i) = \int\!\ldots\!\int f(x_1,\ldots,x_n)\, dx_1 \ldots dx_{i-1}\, dx_{i+1} \ldots dx_n$$

the integral being over the full ranges of all the x_j, $j \neq i$. (Throughout all the following descriptions, in the case of the mass function integrals would be replaced by summations). The conditional density for $\tilde{x}_1, \ldots, \tilde{x}_k$ given $\tilde{x}_{k+1} = x_{k+1}, \ldots, \tilde{x}_n = x_n$ is denoted and defined by

$$f(x_1,\ldots,x_k|x_{k+1},\ldots,x_n) = \frac{f(x_1,\ldots,x_n)}{f(x_{k+1},\ldots,x_n)}$$

In particular, in the case of two random quantities we have

$$f(x_1,x_2) = f(x_1|x_2)f(x_2) = f(x_2|x_1)f(x_1)$$

where, of course, the various f's involved here will have different functional forms. From this, we can deduce that

$$f(x_2|x_1) = \frac{f(x_1|x_2)f(x_2)}{f(x_1)} = \frac{f(x_1|x_2)f(x_2)}{\int f(x_1|x_2)f(x_2)dx_2}$$

where, again, in the discrete case the integral would be replaced by a summation. The above expression is a version of *Bayes' Theorem.*

The kth *moment* of a continuous random quantity $\tilde{x}$ is defined by

$$E(\tilde{x}^k) = \int x^k f(x)dx.$$

The case k = 1 gives the ***mean*** (or ***expectation***), the form $V(\tilde{x}) = E(\tilde{x}^2) - E^2(\tilde{x})$ defines the ***variance*** (whose square root is the ***standard deviation***).

4. Reliability Concepts and Models

In many reliability studies, the key random variables take the form of the "time to failure" of some entity (a unit component, subsystem or system). Let us denote such a random quantity by $\tilde{t}$ and its probability density and cumulative distribution functions by f(t), F(t), respectively. The following derived quantities and functions are key concepts in reliability studies.

Reliability function

$$R(t) = P(\tilde{t}>t) = \int_t f(t)dt = 1 - F(t).$$

Mean time to failure (expected life)

$$E(\tilde{t}) = \int_0^\infty tf(t)dt = \int_0^\infty R(t)dt,$$

the latter equality being derived by integration by parts.

Mean residual life (at t)

$$E(\tilde{t}|\tilde{t}>t) = \frac{1}{R(t)} \int_0^\infty \tau \, f(t+\tau) \, d\tau$$

Interval failure rate

$$FR(t_1,t_2) = \left[\frac{R(t_1) - R(t_2)}{R(t_1)} \right] \cdot \left[\frac{1}{t_2 - t_1} \right]$$

The first term is the conditional probability of failure during the time interval (t_1,t_2), given survival until time t_1. The second term is a "scaling factor" to give the "failure rate per unit time".

Hazard rate (instantaneous failure rate)

If in the previous definition we let $t = t_1$ and $\Delta t = (t_2 - t_1) \rightarrow 0$, we obtain

$$h(t) = \lim_{\Delta t \rightarrow 0} \frac{R(t) - R(t+\Delta t)}{\Delta t \, R(t)}$$

$$= \frac{1}{R(t)} \left[- \frac{dR(t)}{dt} \right] = \frac{f(t)}{R(t)} .$$

If h(t) is an increasing function of t, we refer to an *increasing failure rate* (IFR); if h(t) is a decreasing function of t, we refer to a *decreasing failure rate* (DFR).

Much of the applied probability (mathematical modelling) concern in reliability studies is with different forms for the density f(t) of a "time to failure" random variable $\tilde{t}$. In general, the approach is to select a mathematical form which includes "adjustable labelling parameters", the latter reflecting some summary aspects of the distribution. As we shall see later, "learning from experience" (in the light of data and expert judgements) then reduces to making inferences about the unknown "labelling parameters". To make this labelling explicit in the specification of the probability model, we shall use the conditional density notation, specifically including the labelling parameter as the "conditioning quantity". Examples of commonly used forms include the following (where, in all cases, t is assumed to be a positive continuous random variable).

Exponential

$$f(t|\lambda) = \lambda e^{-\lambda t}, \ \lambda > 0.$$

Here, the interpretation of the labelling parameter λ is clarified by noting that

$$E(\tilde{t}|\lambda) = \lambda^{-1}, \ V(\tilde{t}|\lambda) = \lambda^{-2}, \ R(t) = e^{-\lambda t}, \ h(t) = \lambda.$$

In particular, λ^{-1} is the mean time to failure and λ is the (constant) hazard rate.

Weibull

$$f(t|\alpha,\beta,\theta) = \frac{\beta}{\alpha}\left(\frac{t-\theta}{\alpha}\right)^{\beta-1} exp\left\{-\left(\frac{t-\theta}{\alpha}\right)^{\beta}\right\} \qquad t\geq\theta, \ \alpha,\beta>0$$

Here, the roles of θ, α, β are clarified by noting that

$$E(\tilde{t}|\alpha,\beta,\theta) = \theta + \alpha\,\Gamma\left(\frac{\beta+1}{\beta}\right)$$

$$V(\tilde{t}|\alpha,\beta,\theta) = \alpha^2\left[\Gamma\left(\frac{\beta+2}{\beta}\right) - \Gamma^2\left(\frac{\beta+1}{\beta}\right)\right]$$

$$R(t|\alpha,\beta,\theta) = exp\left\{-\left(\frac{t-\theta}{\alpha}\right)^{\beta}\right\}$$

$$h(t|\alpha,\beta,\theta) = \frac{\beta}{\alpha}\left(\frac{t-\theta}{\alpha}\right)^{\beta-1}$$

where $\Gamma(s) = \int_0^\infty y^{s-1}e^{-y}\, dy$ is the well-known gamma function. For future reference, we note that important functions like the hazard function are *non-linear functions* of the labelling parameters, α, β and θ. We also note that the random variable $\tilde{t}_* = [(\tilde{t}-\theta)/\alpha]^\beta$ has an exponential distribution with "$\lambda = 1$".

Examination of the form of the hazard function reveals that different parameter combinations lead to very different qualitative behaviour (IFR versus DFR).

Normal (Gaussian) and Lognormal

The random variable $\tilde{x}$ is said to have a normal distribution if

$$f(x|\mu,\sigma^2) = \frac{1}{\sqrt{2\pi}\,\sigma}\, exp\left\{-\frac{1}{2\sigma^2}(x-\mu)^2\right\} \qquad -\infty < x < \infty$$

A failure time $\tilde{t}$ is said to have a lognormal distribution if $\tilde{x} = \log \tilde{t}$ has a normal distribution. We shall assume that the reader is thoroughly familiar with the properties of the normal distribution.

Inverse Gaussian

$$f(t|\mu,\lambda) = \left(\frac{\lambda}{2\pi t^3}\right)^{1/2} exp\left\{-\frac{\lambda}{2\mu^2 t}(t-\mu)^2\right\} \qquad \mu,\ \lambda > 0$$

The role of μ, λ is clarified by noting that

$$E(\tilde{t}|\mu,\lambda) = \mu, \quad V(\tilde{t}|\mu,\lambda) = \frac{\mu^3}{\lambda}$$

5. Forms of Data

Suppose available data consist of n precise counts or measurements, t_1, ..., t_n, which are considered *independent*, conditional on the assumption of a specific labelled probability model, $f(t|\theta)$, say, where θ denotes a vector of labelling parameters. We then call $\tilde{t}_1 = t_1, \ldots, \tilde{t}_n = t_n$ a *random sample* from $f(t|\theta)$ and write

$$f(\text{data}|\theta) = f(t_1,\ldots,t_n|\theta) = f(t_1|\theta) \ldots f(t_n|\theta)$$

However, often in reliability studies there is *censoring*, in the sense that a $\tilde{t}$ is not observed precisely, but we merely know, for example, that $\tilde{t} > T$, for some specified T. For instance, suppose a test was undertaken for time T and that, during the test of five items, items 1, 2 and 4 failed, at observed times t_1, t_2, t_4, whereas items 3 and 5 had still not failed at the end of test.

Then

$$\begin{aligned} f(\text{data}|\theta) &= f(t_1|\theta)\, f(t_2|\theta)\, f(t_4|\theta)\, P(\tilde{t}_3 > T|\theta)\, P(\tilde{t}_5 > T|\theta) \\ &= f(t_1|\theta)\, f(t_2|\theta)\, f(t_4|\theta)\, R^2(T|\theta) \end{aligned}$$

Clearly, other forms of censoring (interval or from the left) are possible. Provided we know the form of censoring, $f(\text{data}|\theta)$ can always be written down, but might be an extremely complicated mathematical function of the unknown labelling parameters θ.

6. Statistical Problems

Assuming that a specification of a labelled family of probability distributions for $\tilde{t}$ has been arrived at – either through theoretical considerations (for example, the known form of $h(t)$), or empirically (for example, through exploratory plotting of past data to understand distributional shape) – the problem of "uncertainty" is initially focussed on the unknown labelling parameter θ in $f(\text{data}|\theta)$. Subsequent problems may involve uncertainty about functions of θ (such as the hazard or reliability functions) or uncertainty about future failures given current data. Technically, we are faced with problems of *inference* and *prediction* in the context of parametric statistical models. Before turning to the Bayesian statistical approach to such problems, we shall give a brief review of non-Bayesian approaches.

7. Review of Non-Bayesian Statistical Methods.

Given an assumed *family of probability models*, density or mass functions, as appropriate, and given *data*, typically a combination of precisely observed and censored values, the starting point for statistical analysis is the form $f(data|\theta)$, where θ denotes (the vector of) unknown labelling parameters. From this, we seek to make inferences about θ, or functions of θ, such as $E(\tilde{t}|\theta)$, $V(\tilde{t}|\theta)$, $R(t|\theta)$ or $h(t|\theta)$, and subsequently to make predictions about as yet unobserved future outcomes. Inferences and predictions may focus on providing point or interval estimates of unknown quantities, or on testing the plausibility of an hypothesis, or on comparing the plausibilities of several suggested hypotheses.

Regarding $f(data|\theta)$, there are two quite distinct non-Bayesian approaches to developing statistical methods.

The first is the so-called *Sampling Theory* approach. This seeks to identify procedures which have good "average" or "in the long run" properties in terms of the *sampling distribution*, $f(data|\theta)$, viewed as describing frequency variability over potentially repeated data sets of the same kind. This approach leads to the theory of *unbiased estimation*, *confidence intervals*, *significance tests* and *hypothesis tests* (with associated *Type I* and *Type II* errors).

The second is the so-called *Likelihood Theory* approach. This seeks to base inferences on $f(data|\theta)$ viewed as a "function of θ" for fixed data (namely, that which we have observed). This approach leads to the theory of *maximum likelihood estimation*, *curvature measures of uncertainty* and *likelihood ratio tests*. Sometimes, elements of this and sampling theory are combined, as when long run frequency properties of maximum likelihood estimation and testing recipes are derived.

We shall assume that the reader is familiar with the basic ideas and techniques of these non-Bayesian approaches and so will confine attention here to commenting on what we perceive to be the deficiencies of the approaches.

First, we note that there are *logical difficulties* with the straightforward application of sampling theory concepts. We shall illustrate this by considering simple problems involving unbiased estimates and confidence intervals.

An unbiased estimation problem

Suppose that failures, $\tilde{x}$, occur as a Poisson process with rate θ per unit time. If $\tilde{x} = x$ failures are observed in the first unit time period, find an unbiased estimator of the probability that there will be 0 failures in the next two time periods.

It is easy to see that we require an unbiased estimator of the quantity $exp\{-2\theta\}$, given $\tilde{x} = x$, where

$$f(x|\theta) = \frac{\theta^x e^{-\theta}}{x!} \qquad\qquad x = 0, 1, 2, \ldots$$

If $\hat{\theta}(\tilde{x})$ denotes an unbiased estimator, the definition of the latter requires that

$$\sum_{x=0}^{\infty} \hat{\theta}(x). \frac{\theta^x e^{-\theta}}{x!} = e^{-2\theta},$$

identically in θ. This, in turn, implies that

$$\sum_{x=0}^{\infty} \hat{\theta}(x) \frac{\theta^x}{x!} = e^{-\theta} = \sum_{x=0}^{\infty} (-1)^x \frac{\theta^x}{x!}$$

invoking the well-known series expansion of the exponential function. But, comparing the left-hand and right-hand forms, this implies that

$$\hat{\theta}(x) = (-1)^x$$

Clearly this is a ridiculous estimator! For a start, if x is odd it yields –1, an impossible value for a probability. Moreover, as x gets large (intuitively making it extremely unlikely that there will be 0 failures in the next two time periods) the estimates oscillate between –1 and +1, depending on whether x is odd or even!

A confidence interval problem

Suppose that two measurements $\tilde{x}_1$, $\tilde{x}_2$ are to be made, independent and uniformly distributed over the interval $(\theta-1, \theta+1)$. A 50% confidence interval for θ is required.

Now, it is clear that either both observations lie to the left of θ, or both to the right, or they lie either side (there being two possibilities for the latter, depending on whether $\tilde{x}_1$ is to the left or right). The uniformity of the distribution implies that all four possibilities are equally likely and hence that

$$P\left\{ \left[\min(\tilde{x}_1,\tilde{x}_2),\ \max(\tilde{x}_1,\tilde{x}_2)\right] \ni \theta \right\} = 0.5$$

so that the interval $\left[\min(\tilde{x}_1,\tilde{x}_2),\ \max(\tilde{x}_1,\tilde{x}_2)\right]$ defines a 50% confidence interval for θ.

But, suppose now that we observe $\tilde{x}_1 = x_1$, $\tilde{x}_2 = x_2$ such that $|x_1-x_2| > 1$. A moment's thought will reveal that, in this case, we are sure that θ lies between the min and the max! We are 100% confident!

The fundamental problem with insisting on procedures defined by virtue of their "long run" or "on average" properties is that the very definitions preclude taking specific account of the actual observed data. As we see from these examples, this can lead to illogical implications.

More generally, we can make the following criticisms.

- Neither sampling theory nor likelihood theory is decision oriented; there is no explicit provision for trading off probabilities against cost/benefit considerations.

- Finding forms of unbiased estimators or confidence intervals is typically very difficult in the presence of censored data, or non-standard probability models (even apart from the logical problems indicated above).

- Maximum likelihood estimates often occur on the boundaries of ranges of possible values, or are equal to the minimum or maximum of the sample. In such cases, the estimate is unsatisfactorily "extreme" and curvature measures are useless for providing a measure of uncertainty about the estimate.

- ▹ The unknown parameter θ is typically a vector; but interest may focus on particular individual components, or functions of the components. Generally, sampling theory and likelihood theory do not provide coherent mechanisms for deriving and linking together the uncertainties.

- ▹ It could be argued that *prediction* is among the most important of statistical problems, in the sense that we want to make uncertainty statements about the future given the past. Neither sampling nor likelihood theory directly addresses this problem.

- ▹ Technical problems with sampling theory and likelihood theory are often resolved mathematically by making approximations which are only appropriate when sample sizes are very large. Usually, these involve assumptions of so–called "asymptotic normality". But many (most?) reliability applications are necessarily restricted to small samples.

- ▹ It could be argued that, both because of small samples and the need for a decision orientation, it is important in reliability problems to be able to incorporate expert judgement and experience, as well as data from related but not identical contexts. Again, sampling theory and likelihood theory are not really designed for incorporating such additional information.

8. Desiderata for Decision-Oriented Statistical Methodology

In contrast to the critical points made in the last section, we can identify desirable aspects that we would like a statistical approach to reliability problems to encompass.

- ▹ The ability to incorporate expert judgements into the analysis.

- ▹ The assurance that expert judgements, or data inputs from other sources, are combined logically with current data in quantifying uncertainties.
- ▹ An approach to uncertainty that facilitates communication, decision making, and analysis of the sensitivity of results to assumptions.

- ▹ A unified approach, which places no reliance on special tricks or magic insights in tackling invididual problems.

▷ An approach which exploits and is well adapted to modern computing facilities.

We shall begin our study of such a methodology by first focussing on the problem of "optimal decision making".

9. Decision-Making

There is no neutral, purely mathematical solution to the problem of choosing an 'optimal' action, but one approach to the subject is to start from primitive concepts and try to *deduce* the form of structure required and optimal approach implied.

To do this, we could make a list of axioms (or 'self-evident' postulates) and then attempt to deduce their implications. We shall not attempt a detailed rigorous approach of this kind, but rather attempt to give the 'flavour' of such an approach and such arguments by discussing part of a particular, formal axiom system which lays down postulates for 'rational preferences' among consequences. In the next section, we shall give an informal discussion of the way in which the notion of 'rational degree of belief' can be analysed.

Suppose that $\mathcal{C}$ is the set of all 'consequences' that can arise in a particular decision problem. Now let $\mathcal{P}$ denote the set of all probability distributions over $\mathcal{C}$ so that $\mathcal{P}$ denotes the set of all possible uncertainties regarding consequences. We now assume that the decision-maker has preferences among the elements of $\mathcal{P}$ and we write $P_1 < P_2$ to denote that P_2 is strictly preferred to P_1, $P_1 \leq P_2$ to denote that P_1 is not strictly preferred to P_2 and $P_1 \sim P_2$ to denote that neither of P_1, P_2 is preferred to the other.

It is assumed that preferences obey the following two axioms.

A1. If P_1, P_2 are elements of $\mathcal{P}$ then either $P_1 < P_2$ or $P_2 < P_1$ or $P_1 \sim P_2$.
A2. If P_1, P_2, P_3 are elements of $\mathcal{P}$ such that $P_1 \leq P_2$ and $P_2 \leq P_3$ then $P_1 \leq P_3$.

In other words, we assume that any two uncertain situations regarding consequences can be compared and that preferences are transitive. Now let us make the further assumption

A3. If P_1, P_2, P_3 are elements of $\mathcal{P}$ and α is any number such that $0<\alpha<1$, then $P_1 < P_2$ if and only if $\alpha P_1 + (1-\alpha)P_3 < \alpha P_2 + (1-\alpha)P_3$.

This assumption formalizes the intuitive idea that if two situations of uncertainty about consequences have a common component the comparison of the two should not depend on this component. We further assume that:

A4. If P_1, P_2, P_3 are elements of $\mathcal{P}$ such that $P_1 < P_2 < P_3$, then there exist numbers α, β $(0<\alpha<1,\ 0<\beta<1)$ such that

$$P_2 < \alpha P_3 + (1-\alpha)P_1 \text{ and } P_2 > \beta P_3 + (1-\beta)P_1$$

This assumption formalizes the idea that 'absolute heaven' and 'absolute hell' are not perceived to be among the consequences. For example,

$$P_1 < P_2 < P_3 \text{ and } P_2 < \alpha P_3 + (1-\alpha)P_1$$

says that, although $P_2 > P_1$, there exists some $1-\alpha$ (however tiny) so that the mixture $\alpha P_3 + (1-\alpha)P_1$ is still preferred to P_2. If P_1 were an 'absolute hell', this would not be so. Similar remarks apply to the other inequality.

These assumptions, and the discussion of them, should serve to indicate how one sets about formalizing intuitive notions about 'rational' or 'coherent' preferences. What we then would do in a detailed analysis is to attempt to deduce from such assumptions the form that rational decision-making procedures should take.

The answer that emerges is that rational decision-making requires us to
(a) assume the existence of a utility function,
(b) act so as to maximize expected utility,
where 'rational' means conforming to the axioms laid down for preferences.
In short, axiom systems of this kind (and there are many variations in the precise list of axioms chosen) lead to the conclusion that the so-called Bayesian approach to decision-making is required if we are to act in conformity with the axiom system.

A detailed account of a system like the one outlined above can be found in M.H. DeGroot, *Optimal Statistical Decisions* (Chapter 7). A more informal treatment is given in D.V. Lindley, *Making Decisions.*

10. Degrees of Belief as Probabilities

We have assumed in Section 2 that 'degrees of belief' can be represented by following the rules of the probability calculus. Some readers may find this entirely reasonable, without the need for further discussion: others may require at least an outline argument in support of this assumption.

One such argument is the following. Suppose that an individual, you, for example, is faced with a situation of uncertainty involving an event E whose outcome (E or not–E) is unknown. In any practical situation of interest to you, you will have some feeling, let us call it 'degree of belief', regarding the relative 'plausibilities' of E or ~E (not–E) occurring. It is this feeling which we wish first to give a quantitative form, and then to show that such quantities must (under a further condition, which we shall make clear) obey the rules of probability.

First, we note that an intuitive operational scheme for converting a degree of belief into a number is the following.

Consider a gamble which offers you £S if E occurs and £0 if ~E occurs. What sum of money, £C, say, would leave you *indifferent* between possession of C for certain and entering the gamble (just once)?

Clearly, if C is 'very small' (as a fraction of S) you would prefer to take the gamble: if, on the other hand, C were large (as a fraction of S), you would prefer to accept C for sure rather than enter the gamble. We shall assume that, as you think through reactions to a 'small' C, getting larger, or a 'large' C, getting smaller, you will perceive that for some intermediate value of C you are indifferent between C for sure and the gamble. Let us call this intermediate value £C^*. (Of course, in practice there will be a 'fuzzy' interval of indifference, rather than a sharp value, but we are used to 'idealizing' a little with *all* forms of measurement; for example, we base much science and technology on the assumptions that bodies have precise 'length' or 'temperature', although in practice these can only be measured up to some 'fuzzy' interval).

Now *define* your (revealed) degree of belief, p, in E to be the number such that

$$C^{*} = pS$$

We note first that this corresponds to our intuitive understanding; low degrees of belief make C^{*} small: high degrees of belief make C^{*} large.

In order to ensure that you are *honest* in your revealed indifference value C^{*}, we can add the following condition.

Suppose that when you state C^{*} you do not know whether you will be called upon *to gamble* (in accordance with your stated value you should be willing to pay C^{*} to enter a gamble with E leading to £S and ~E leading £0), or *to act as bookmaker* (offering this gamble for a stake of C^{*}). This implies that you must state a C^{*} which leaves you indifferent in both the contexts.

Having now indicated a way in which an honest, quantitative degree of belief for an event E can be arrived at, we wish to examine the ways in which degrees of beliefs for different events relate to each other.

Consider the following situation. The collection of events E_1, E_2, ..., E_N are exclusive and exhaustive and using the above procedure you are to specify degrees of belief p_1, p_2, ..., p_N in the individual events. After you have chosen p_1, p_2, ..., p_N, an 'opponent' will be free to choose amounts S_1, S_2, ..., S_N on the assumption that you will be willing to pay entry fees, p_1S_1, p_2S_2, ..., p_NS_N to a gamble in which you will receive C_1 if event E_1 occurs, C_2 if E_2 occurs, and so on.

How must the p_i's 'fit together', or 'cohere', if you are to avoid the situation where the opponent can choose the S_i's in such a way that he or she will certainly win? Any specification of the p_i's such that you could certainly lose, we shall call ***irrational***. Our question then becomes 'what rules must ***rational*** degrees of belief obey?'

We note first that $0 \leq p_i \leq 1$, for all $i = 1, ..., N$, since, recalling the definition of degree of belief, any choice outside this range would enable the opponent to choose his or her role (as gambler, or bookmaker) so as to ensure that he won. Specifically, a choice of $p_i > 1$ would imply that you were willing to pay > S_i to enter a gamble with maximum prize of S_i; a choice of $p_i < 0$ would imply that you were willing to *pay* your opponent to enter a gamble which would either yield him or her (from you!) 0 or S_i.

Secondly, we note that *your* 'gains', G_i (which may, of course, be negative), satisfy the system of linear equations

$$G_i = S_i - (p_1S_1 + \dots + p_NS_N)$$

if E_i occurs, i = 1, ..., N. (The total entry fee is Σp_jS_j, the 'payoff' is S_i if E_i occurs).

Written out in full, the system is

$$\begin{bmatrix} G_1 \\ G_2 \\ G_3 \\ \vdots \\ G_N \end{bmatrix} = \begin{bmatrix} 1-p_1 & -p_2 & p_N \\ -p_1 & 1-p_2 & -p_N \\ \vdots & & \vdots \\ -p_1 & -p_2 & 1-p_N \end{bmatrix} \cdot \begin{bmatrix} S_1 \\ S_2 \\ \vdots \\ S_N \end{bmatrix}$$

and the problem can be reformulated as follows. The opponent is given the p_i's and wishes to choose the S_i's in order to give G_i's values which he or she (the opponent) desires. In particular, he or she would wish to fix negative G_i's (remember they are *your* 'gains'). For a given set of G_i's and p_i's, can he or she solve the equations and find suitable S_i's? A *rational* set of p_i's is one that prevents the opponent being able to solve these equations. We now recall, from linear algebra, than an explicit solution *can* be found if the matrix (defined by the p_i's) can be inverted. We also recall that this is *not possible if the determinant is zero.*

An easy calculation shows that the determinant is equal to $1 - (p_1+\dots+p_N)$, so that a collection of rational degrees of belief for the exhaustive and exclusive events $E_1, \dots, E_N$ must satisfy

$$p_1 + \dots + p_N = 1$$

We see therefore that rational degrees of belief satisfy the basic rules of probability: they have values in the range from 0 to 1; they satisfy the addition axiom.

To complete an outline justification of regarding degrees of belief as probabilities, we must consider the concept of 'conditional' degrees of belief, or the way in which degrees of belief change when we acquire new information.

Let us consider two events E' and E", and let

π_1 = p(E'|E")	*denote degrees*	E' given the occurrence of E"
π_2 = p(E' and E")	*of belief*	the occurrence of both E' and E"
π_3 = p(E")	*of*	the occurrence of E".

You are now to gamble (the outcomes being defined in terms of E', E") on the understanding that if E" does not occur then bets involving E' given E" are 'called-off' (i.e. the entry fee is returned). How should π_1, π_2, π_3 relate to one another if you are to avoid an opponent fixing stakes which lead you to certainly lose?

Suppose that we consider the three contingencies E' given E", E' and E", E", respectively, with amounts S_1, S_2, S_3 as 'pay-offs'. It is easily seen that the following table of 'gains' (for you) arises from the stated outcomes.

Actual outcome	*Gain*
E' and E"	$G_1 = (1-\pi_1)S_1 + (1-\pi_2)S_2 + (1-\pi_3)S_3$
~ E' and E"	$G_2 = -\pi_1 S_1 - \pi_2 S_2 + (1-\pi_3)S_3$
~E"	$G_3 = -\pi_2 S_2 - \pi_3 S_3$

This may be rewritten in the form

$$\begin{bmatrix} G_1 \\ G_2 \\ G_3 \end{bmatrix} = \begin{bmatrix} 1-\pi_1 & 1-\pi_2 & 1-\pi_3 \\ -\pi_1 & -\pi_2 & 1-\pi_3 \\ 0 & -\pi_2 & -\pi_3 \end{bmatrix} \cdot \begin{bmatrix} S_1 \\ S_2 \\ S_3 \end{bmatrix}$$

and the argument we used previously tells us that the π_i's must be chosen so that the determinant is zero if irrationality is to be avoided.

In this case, the determinant is equal to $\pi_2 - \pi_1\pi_3$ and so the degrees of belief must satisfy

$$p(E'|E") = \frac{p(E' \text{and } E")}{p(E")}$$

This establishes that rational, conditional degrees of belief satisfy the usual definition of conditional probability.

It could be objected that if the sums S_i used in the above arguments are large then the procedure for defining degrees of belief would be invalidated by the phenomenon of risk aversion. This is true, and so one must understand the above argument as either conducted with 'small' S_i, so that utility is (approximately) linear, or with the S_i defined on a preassessed utility scale.

11. Bayesian Statistical Philosophy

As we have noted, statisticians typically try to interpret data via so-called statistical models. These involve assumed mathematical relationships between observed data and unobserved – and thus unknown – underlying aspects of the problem under study. Uncertain aspects of these relationships are further described by chance mechanisms – for example, the "noise" inherent in a measuring device, or the sampling variation that may distort observed frequencies in a survey.

In Section 7, we noted that conventional statistical methods take the posited uncertainty relationships between data and unknowns (the latter often referred to as parameters) and try to provide recipes for answering questions about the parameters. For example: How to provide a single number estimate of the unknown true value? How to test (yes or no) whether the true value is equal to some hypothesized value? How to give an interval of uncertainty (a so-called confidence interval) so that "in the long run" the true value would be found to be in the interval very often (for example, 90% of the time)?

Bayesian statistical methods do not follow this path of inventing recipes (estimates, tests, confidence intervals) for each new one-off problem encountered. Instead, the problem of dealing with uncertainty in the context of an ongoing process of data collection is viewed simply as:

REVISING CURRENT BELIEFS IN THE LIGHT OF NEW INFORMATION.

Put this way, statistics is not seen as a "bag of tricks", far removed from everyday experience. Rather, it is viewed simply as a formal scientific way of dealing with life's ubiquitous problem of LEARNING FROM EXPERIENCE.

Given this philospohy, there are two questions to be answered. What is the appropriate scientific (mathematical) way to represent beliefs? What is the appropriate scientific (mathematical) way to update current beliefs into revised beliefs when new data is obtained?

In the previous two sections, we have argued that the answer to the first question is that, if rational decision-making is the ultimate goal, then belief about unknowns should be represented in the form of probability descriptions. Queries about unknowns are then always answerable in terms of statements about relative odds or chances (directly quantified and clearly "act-onable").

The answer to the second question is more technical, but easily explained. As a preliminary, we note that we have three "ingredients" which need fitting together logically: first, the statisticians's model of how the data collection mechanism relates to the unknown parameters; secondly, the decision-maker's beliefs about the unknown parameters before seeing the data; thirdly, the decision-maker's beliefs after seeing the data. In statistical jargon, these are referred to, respectively, as the likelihood function, the prior density and the posterior density.

Let us denote these three terms by P(data|parameters), P(parameters) and P(parameters|data), respectively. Read the first of these as "how likely are the data to arise given some specified values of the unknown parameters" (the likelihood); the second as "how much belief do I attach to these specified values of the unknown parameters before (i.e. prior to) observing the data"; the third as "how much belief do I attach to these specified values of the unknown parameters after (i.e. posterior to) observing the data". Finally, let us denote by P(data) the chance, averaged over all the likelihoods, of obtaining the observed data.

Then the key to all Bayesian statistical methods is provided by Bayes' theorem which states that

$$\text{P(parameters|data)} = \frac{\text{P(data|parameters)}}{\text{(data)}} \times \text{P(parameters)}$$

In more mathematical terms, we have

$$p(\theta|\text{data}) = \frac{p(\text{data}|\theta)p(\theta)}{\int p(\text{data}|\theta)p(\theta)d\theta}$$

If the first term on the right-hand side is renamed the standardized likelihood, we can state Bayes' theorem in words as:

$$\text{POSTERIOR BELIEFS} = \text{STANDARDIZED LIKELIHOOD} \times \text{PRIOR BELIEFS}$$

How does the formula work? It's simple and very commonsensical. Values of the unknown parameters which give rise to large values of the standardized likelihood will lead to higher posterior beliefs than values of the unknown parameters which give rise to small values of the standardized likelihood. And what does the latter measure? Well, high values correspond precisely to parameter values which are "likely" to have led to the data we actually observed. So, we increase our beliefs in parameters values which are highly compatible with the data we've seen; we correspondingly decrease our beliefs in parameter values which are unlikely to have resulted in the data actually observed.

And what happens next? That is, when still further new data is observed? Again, it's very simple. Basically, today's posterior becomes tomorrow's prior. In other words: our new "current" beliefs are P(parameters|previous data); these will be revised via Bayes' theorem to

P(parameters|previous and new data)

and so it goes on. Bayes' theorem is a perfectly natural tool for successive revision of beliefs as each piece of new data comes in. We simply "learn from experience".

12. A Simple Illustration of Bayesian Learning

Let us illustrate some features of the learning process when the unknowns consists of a single parameter. In this case, beliefs are represented by a curve and changes in such curves can easily be displayed graphically. Think of the unknown as a breaking strength, or absorption coefficient, or some other physical attribute of a new material. Now suppose that we consider two illustrative forms of prior beliefs. One (B) corresponds to an individual who is rather ignorant about material of this type and hence vague about the value of the unknown parameter. In fact, his best guess is around a value of 400, but he regards all values in a range from about 200 to 650 as possible. The other (A) is rather knowledgeable about material of this type and thus has quite tightly defined beliefs. His best guess is around 500, but he is practically sure that the true value lies in the range 450 to 550. Belief curves describing the prior beliefs of A and B are shown in Figure 1.

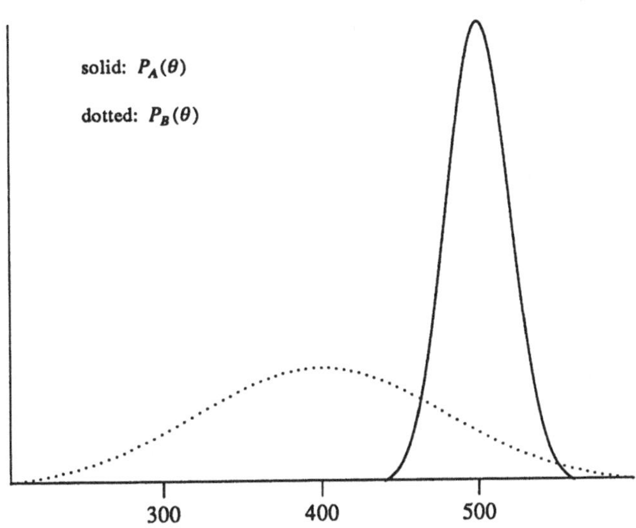

Figure 1. Prior densities for A and B.

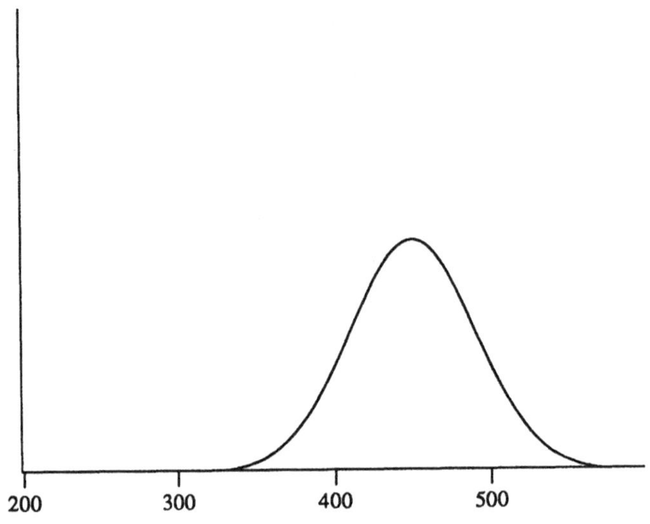

Figure 2. Shape of likelihood for 1 observation.

Now suppose that a single measurement of the quantity of interest is made available and gives a value of 450 with a standard error of measurement of about 50. The likelihood to which this gives rise is shown in Figure 2.

Let's recall, informally, what Bayes' theorem tells us to do in order to see how the prior beliefs of A and B should be revised in the light of the data to give updated (posterior to the data) beliefs. The theorem says:

Posterior beliefs are proportional to likelihood $\times$ prior beliefs.

Mathematically, we simply multiply together the curve shown in Figure 2 with each of the separate curves shown in Figure 1 and then standardize. The resulting curves of posterior belief for A and B are shown in Figure 3 (cf. Badoux section 2).

We see immediately that the single observation has changed B's beliefs quite dramatically. Whereas previously his best guess was around 400, with a range of 200 to 650, his beliefs are now centered around about 430 (much nearer to the observed measurement) with a range of 350 to 500. Having been very vague prior to seeing data, the data has had the two-fold effect of sharpening his beliefs (tighter range) and influencing where his beliefs are centered (now nearer to 450 than 400).

A's beliefs have also sharpened up and recentered themselves, but the change is much less marked. The new data has, of course, added information. But, since A was already quite knowledgeable, the amount of new information is not such a dominant factor as was the case with B. A's best guess has shifted from 500 to about 485; the range of his beliefs has shifted from 450–550 to 440–525 – pulled downwards by the observed measurement of 450.

We have so far seen how a relatively small amount of data (in fact, a single observation) influences the prior to posterior learning process for two rather different prior belief profiles. Suppose now that instead of just a single observation we made dozens, or even hundreds, of measurements and averaged them. The standard error of measurement would then be much reduced (typically by a factor equal to the square root of the number of measurements taken; for example, by a factor of 10 if we average 100 measurements). In graphical terms, this results in a tight, highly peaked likelihood, like that based on 100 measurements shown in Figure 4 (with the two prior curves for A and B, redrawn to be on the same scale).

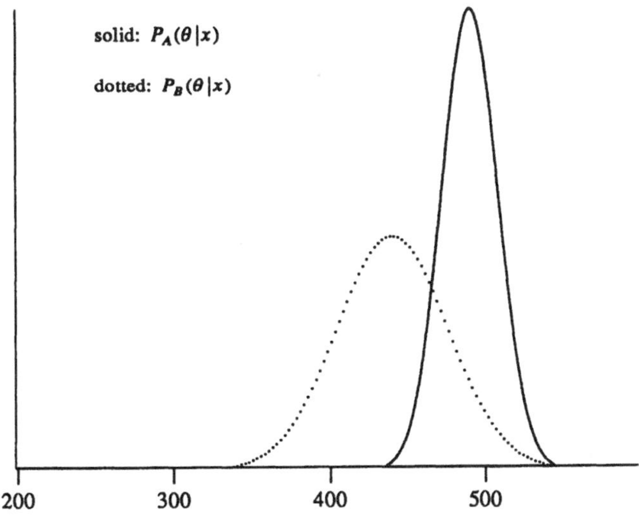

Figure 3. Posterior densities for A and B after 1 observation.

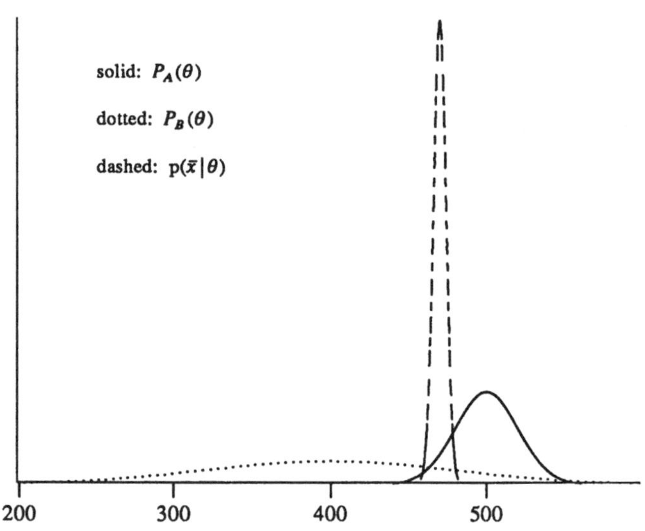

Figure 4. Priors for A and B with shape of likelihood from 100 observations.

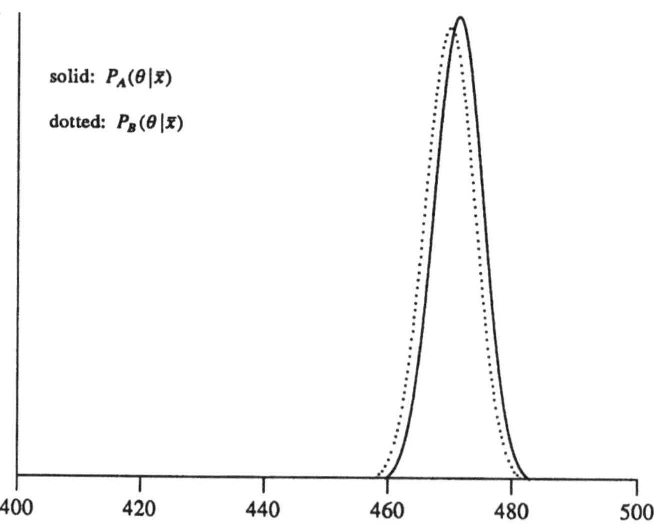

Figure 5. Posteriors for A and B after 100 observations.

Recalling that Bayes' theorem tells us to multiply together the prior and likelihood curves in order to get forms proportional to the posterior likelihood curves for A and B, it is easy to see from Figure 4, what will happen. The sharply peaked likelihood function dominates and leads to the two posterior belief curves shown in Figure 5.

We observe from this latter picture that the message in the data has completely overwhelmed the prior beliefs of both A and B, whose beliefs now more or less coincide and are shaped according to the standardized likelihood.

The flavour and mechanism of Bayesian learning emerges clearly from Figures 1–5. Different individuals will have different prior beliefs, depending on their knowledge and experience. Data – in the form of measurements or counts – gives rise to a likelihood when linked to unknowns via a statistical model. Bayes' theorem asserts that the rational way to modify prior beliefs in the light of data is to multiply prior and standardized likelihood curves together to produce a posterior curve. For relatively small amounts of data, posterior beliefs tend to be a compromise between prior beliefs and the message from the data. But large amounts of data will typically overwhelm prior beliefs (unless the latter are very extreme) and force posterior beliefs to follow very closely the message in the data (as revealed through the statistical model used). In the latter case, an "objective consensus" may emerge.

13. Bayesian Approaches to Typical Statistical Questions

Point Estimation

One of the most commonly posed problems is that of providing, on the basis of observed data, an estimate of an unknown parameter.

As we have seen, given a model involving an unknown parameter θ, and given data x, the Bayesian statistician can calculate the posterior density $p(\theta|x)$, corresponding to any particular prior density specification for θ. In a sense, given x and the specified likelihood and prior density, the description of the form of $p(\theta|x)$, either analytically as a mathematical function, or graphically, provides a 'complete picture' of what is now believed about the unknown parameter θ. If we are asked to give a single (point) estimate of θ, the question for a Bayesian statistician becomes

'What single number *best summarizes* beliefs, as represented by $p(\theta|x)$?'

Of course, the phrase, 'best summary' is, as it stands, undefined: we must ask 'best with respect to what criterion?' But as soon as we begin to pose such questions, we are led towards the idea that in order to judge whether a method of choosing an estimate is sensible or not we must know something about the practical problem under study, and the actual consequences of the various discrepancies between 'estimate' and 'true value' that might arise. When viewed in this light, statistical estimation is seen to be a special kind of 'decision', whose potential consequences must be quantified before an 'optimal decision' can be made (or, in this case, 'best estimate' provided).

A detailed presentation of the decision–theoretic approach to estimation is given in the book by DeGroot, cited in Section 9. We shall just consider three of the possible summaries of the posterior density that might be thought *intuitively* appealing

(i) *Mode.* The mode, θ^*, of the posterior density is defined by

$$p(\theta^*|x) = \sup p(\theta|x)$$

and is typically unique. The estimate θ^* is the 'most likely' value of θ and is (approximately) equal to the maximum likelihoood estimate in situations where the prior density is (approximately) specified to be constant.

(ii) *Median.* The median of the posterior density, $\tilde{\theta}$, is defined by

$$\int_{-\infty}^{\tilde{\theta}} p(\theta|x)d\theta = \int_{\tilde{\theta}}^{\infty} p(\theta|x)d\theta$$

and is typically unique. It is the point such that we have equal beliefs that the true θ lies above or below the value.

(iii) *Mean.* The mean of the posterior density, $\hat{\theta}$, is defined by

$$\hat{\theta} = \int_{-\infty}^{\infty} \theta p(\theta|x) d\theta$$

and is unique.

Interval Estimation

It is clear that whatever form of point estimate is chosen it provides a poor summary of the complete posterior density, $p(\theta|x)$. A half-way house between the, perhaps complicated, description of the complete density and the over-simplification of the point estimate is the idea of a *credible interval.*

Given a posterior density function $p(\theta|x)$, we quote two values of θ, a and b ($a<b$), such that the posterior probability of θ lying in the interval from a to b is equal to some specified value (say, 90, 95, or 99%, or whatever is appropriate for the problem under study). *More precisely, we say that, given $p(\theta|x)$, the interval (a,b) is a $100(1-\alpha)\%$ posterior credible interval of θ if*

$$\int_{a}^{b} p(\theta|x) d\theta = 1-\alpha \qquad (0 \leq \alpha \leq 1)$$

When we use values 0.1, 0.05 or 0.01 for α, we speak of 90, 95, or 99% credible intervals for θ.

It will be seen that a 'credible interval' is formally identified with a 'probability interval', where the probability interval is taken with respect to the posterior distribution $p(\theta|x)$.

In general, we can find many pairs of values (a,b) which provide a $100(1-\alpha)\%$ credible interval, for a specified α. This observation motivates a more refined notion of credible interval. An interval (a,b) is said to be a 100 $(1-\alpha)\%$ *highest posterior density interval* if

(i) (a,b) is a $100(1-\alpha)\%$ credible interval;

(ii) for all $\theta' \in (a,b)$ and $\theta'' \notin (a,b)$, $p(\theta'|x) \geq p(\theta''|x)$.

The extra condition (ii) requires that no value of θ included in the interval (a,b) should have an ordinate, $p(\theta|x)$, of the posterior density lower than any value of θ excluded from (a,b). Clearly, this produces the shortest possible credible interval for a given α, and hence the 'most informative' interval estimate summary of $p(\theta|x)$.

It is important to distinguish the credible interval approach from the superficially similar approach to interval estimation based on confidence intervals. The former uses $p(\theta|x)$ and refers directly to the *probability of θ lying in a particular interval.* The latter considers intervals with random endpoints, $a(\tilde{x})$ and $b(\tilde{x})$, which have the property that, in terms of the distribution $f(x|\theta)$, *the probability of the interval containing θ has the value* $1-\alpha$, for suitably specified α. The particular interval $[a(x), b(x)]$ which obtains when $\tilde{x} = x$ is observed is then referred to as a $100(1-\alpha)\%$ confidence interval.

Significance Testing

In many situations modelled by a family of densities, $f(x|\theta)$ with unknown θ, it is of interest to examine whether the data is or is not 'compatible' with a *particular* value of the unknown parameter, θ_0 say. If we consider H_0: $\theta = \theta_0$ as a basic hypothesis of interest (usually called the *null hypothesis*), we are led to seek a procedure for 'rejecting' or 'not rejecting' such an hypothesis.

The *significance testing* approach is very familiar in a non-Bayesian framework. One way of looking at such non-Bayesian significance tests in the case of parametric models, $f(x|\theta)$, is to note that we 'reject' H_0: $\theta = \theta_0$ at the $100\alpha\%$ level if θ_0 does not lie in a (sensible) $100(1-\alpha)\%$ confidence interval.

Proceeding analogously, a possible *Bayesian significance test* is provided by rejecting H_0: $\theta = \theta_0$ 'at the $100\alpha\%$ level' if θ_0 lies outside the $100(1-\alpha)\%$ *highest posterior density interval.*

Prediction

If we consider 'predicting' a future observation or observations, x say, on the basis of previous data y, the Bayesian statistician will seek to derive a distribution of belief for x, given y. Probabilistically, we are led to consider $\pi(x|y)$, the so-called *predictive density* for x, given y.

Typically, we shall not have a model which specifies this density directly. Rather, we will usually have a probability model for x, $g(x|\theta)$, depending on an unknown parameter θ, which itself appears in the model assumed for y. If $p(\theta|y)$ is the posterior density for θ, given the previous data y, and if x and y are independent, given θ, we may obtain the predictive density for x from the formula

$$\pi(x|y) = \int_{\Theta} g(x|\theta)p(\theta|y)\, d\theta.$$

If a 'single-figure' prediction is required, we simply choose a point estimate summary of the density $\pi(x|y)$ using, for example, mode, median or mean. If a predictive interval is required, we can use the idea of a credible interval or a 'highest predictive density interval', an obvious modification of earlier ideas applied to the density $\pi(x|y)$.

Summarizing Data: Sufficient Statistics

We consider the problem of summarizing a set of observations in the form of a small number of summary statistics without losing relevant information. In non-Bayesian statistics, this leads to the familiar notion of *sufficient statistics.*

In order to re-examine this problem from a Bayesian point of view, we can argue as follows. Given a likelihood defined by $f(x|\theta)$ and a prior density $p(\theta)$, the Bayesian approach tells us to combine these, using Bayes' Theorem, and thus to determine $p(\theta|x)$, the posterior density given *all the data, x.* Suppose now that t(x) is some summary of the data x; for example, we might have $x = (x_1, ..., x_n)$ and $t(x) = n^{-1}\Sigma\, x_i$, or $t(x) = \min\{x_1, ..., x_n\}$, etc. Since $p(\theta|x)$ is the 'complete' representation of current beliefs about θ, given x, the summary t(x) can only be said 'not to have lost any relevant information' if $p(\theta|t(x))$ is equal to $p(\theta|x)$. If, given $f(x|\theta)$, statisticians with different specifications of $p(\theta)$ are to agree that t(x) constitutes an acceptable sufficient summary we should require that

$$p(\theta|t(x)) = p(\theta|x) \text{ for all } p(\theta).$$

If this condition holds, we say that t(x) is a *Bayes' sufficient statistic.*

This definition of a sufficient statistic appears to be entirely different from the more familiar (non–Bayesian) definition. In fact, however, the two definitions can be shown to be equivalent. Provided, therefore, that the form of probability model $f(x|\theta)$ is agreed, both Bayesian and non–Bayesian statisticans will base their analyses on the same data summary (the sufficient statistics).

The Likelihood Principle

Suppose an investigation consists of a sequence of independent experiments, each having a chance θ of resulting in a success and $1-\theta$ of resulting in a failure. Suppose further that it is reported that a total of n experiments were performed and y successes were obtained.

If we denote the data by $x = (n,y)$, do we have sufficient information to write down a probability model relating x to θ?

The answer, of course, is *no*, since we are not told what method of experimentation (or sampling) was employed. For example, if n were fixed in advance and we simply observed y then we would have the Standard Binomial probability distribution with

$$f(y|\theta,n) = \binom{n}{y} \theta^y(1-\theta)^{n-y}, \qquad y = 0, 1, \ldots, n$$

If, on the other hand, we had decided to fix y in advance and then observe n (that is, to continue experimenting until y successes were achieved and then to note how many experiments, n, had been required) we would instead have the Negative–Binomial distribution with

$$f(n|\theta,y) = \binom{n-1}{y-1} \theta^y(1-\theta)^{n-y}, \qquad n = y, y+1, y+2, \ldots$$

There are, of course, many other sampling rules that might have been employed (for example, 'continue experimenting until lunchtime, and then stop, fixing neither n or y'), but the point to which we wish to draw attention can be illustrated using just the Binomial and Negative–Binomial forms.

The basic question is as follows: given the data alone, that is, without being told the method of sampling, can we make standard inference statements about θ? For example, can we provide point or interval estimates?

The answer to this is that the Bayesian approach *can* provide inferences without knowing whether the Binomial or Negative–Binomial distribution is appropriate, whereas many non–Bayesian procedures *cannot*.

To see this, let us make the assumption that prior beliefs about θ are not influenced by the form of sampling employed, so that we can specify $p(\theta)$ independently of the form of likelihood.

For the ***Binomial*** assumption, Bayes' Theorem gives

$$
\begin{aligned}
p(\theta|x) &= p(\theta|n,y) \propto f(y|\theta,n)p(\theta) \\
&\propto \binom{n}{y}\theta^{y}(1-\theta)^{n-y}\; p(\theta) \\
&\propto \theta^{y}(1-\theta)^{n-y}p(\theta),
\end{aligned}
$$

since $\binom{n}{y}$ does not involve θ, and thus

$$
p(\theta|x) = \frac{\theta^{y}(1-\theta)^{n-y}p(\theta)}{\int_0^1 \theta^{y}(1-\theta)^{n-y}p(\theta)\; d\theta}.
$$

For the ***Negative–Binomial*** assumption, Bayes' Theorem gives

$$
\begin{aligned}
p(\theta|x) &= p(\theta|n,y) \propto f(n|\theta,y)\; p(\theta) \\
&\propto \binom{n-1}{y-1}\; \theta^{y}(1-\theta)^{n-y}p(\theta) \\
&\propto \theta^{y}(1-\theta)^{n-y}p(\theta),
\end{aligned}
$$

since $\binom{n-1}{y-1}$ does not involve θ, and thus

$$p(\theta|x) = \frac{\theta^y(1-\theta)^{n-y}p(\theta)}{\int_0^1 \theta^y(1-\theta)^{n-y}p(\theta)\, d\theta}.$$

The posterior density $p(\theta|x)$ *is thus seen to be identical in the two cases.*

However, if we consider non-Bayesian procedures, such as minimum variance unbiased estimation (MVUE), confidence intervals or significance tests, the precise forms of these differ according as we make the Binomial or Negative-Binomial assumption. For example, under the Binomial assumption $\tilde{y}/n$ is a MVUE, whereas under the Negative-Binomial assumption $(\tilde{y}-1)/(n-1)$ is the MVUE

The Binomial/Negative-Binomial example is a special case of the following situation. We have two different probability models

$$f_1(x|\theta) = c_1(x)g(x,\theta), \qquad f_2(x|\theta) = c_2(x)g(x,\theta)$$

with two different ranges of possible values of x. The part of the density depending on θ is the same in the two cases. The proportional form of Bayes' Theorem makes clear that, given x and $p(\theta)$, the form of $p(\theta|x)$ will be the same in both cases, since it does not require the explicit form of $c_i(x)$, $i=1,2$.

Another way of interpreting this is to note that $g(x,\theta)$ takes into account only the *actual x* observed, whereas the forms of $c_i(.)$, $i=1,2$, reflect the range of *all possible x* that might have been obtained.

The ***likelihood principle*** suggests that inferences about θ should be based only upon $g(x,\theta)$, not taking into account $c_i(x)$, $i = 1,2$. Bayesian procedures (and others such as maximum likelihood) may be said to *obey* the likelihood principle; procedures like MVUE are said to *violate* the likelihood principle.

14. Assessment of Prior Densities

The specification of the prior density component of Bayes' Theorem depends on individual experience and beliefs. This poses the very real problem of how to convert such experiences and beliefs into the form of a probability density function $p(\theta)$. This problem has been exhaustively studied by both statisticians and psychologists, and many theoretical and experimental papers have been published reporting the results of such studies.

Among the many possible approaches, the following are, perhaps, the two most important.

(i) *Smoothing of historical data*

Suppose that a manufacturer is uncertain about the proportion, θ, of defective items that will result from a new production process that is to be introduced, but has available a histogram showing the relative frequencies with which the proportion of defectives fell in various intervals when a number of very similar previous production processes were introduced. Figure 6 shows such a histogram, together with a smooth curve, which, suitably normalized so as to contain total area 1, might well reasonably reflect, in the form of a probability density $p(\theta)$, the manufacturer's prior beliefs.

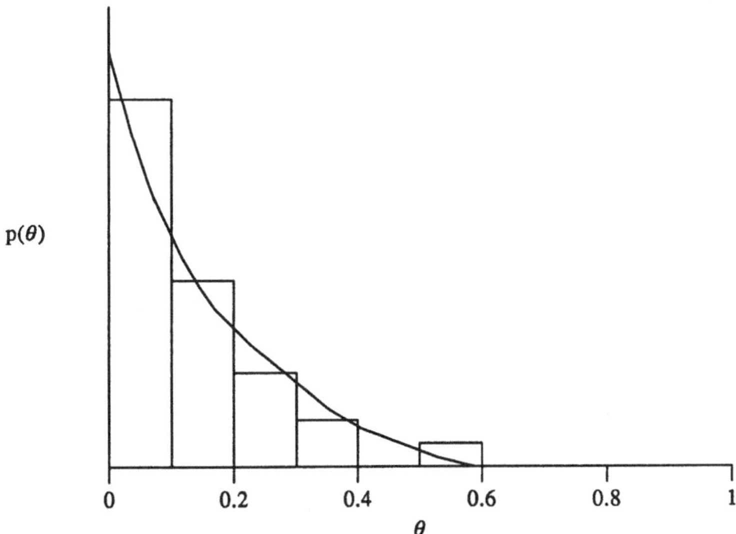

Figure 6. $p(\theta)$ as smoothing of historical data.

The technique is very straightforward: we simply seek to produce a smooth density which reflects the form of an historical frequency distribution. Even here, however, we cannot escape the subjective, judgemental nature of the specification, since we must *assess* the relevance and homogeneity of the historical data in relation to our current problem.

(ii) *Judgemental curve fitting*

In the absence of sufficient, relevant past data to permit the use of (i), we are forced to try to elicit beliefs directly by a process of (self–) interrogation.

Consider, for example, a manufacturer introducing an entirely new type of production process, who does not feel that historical data regarding the proportions of defectives resulting from other types of process is directly relevant to his or her new process.

The following kind of procedure can be used to elicit directly his or her beliefs about θ, the proportion of defectives that will occur under the new process.

(a) The manufacturer is asked to give upper and lower limits between which, in his or her opinion, the proportion θ will be. In practice, we might ask for values such that there is only, according to the manufacturer, a one in a hundred chance of each of these limits being exceeded. The values specified would then give, approximately, the first and ninety–ninth percentiles of his or her distribution.

(b) We then ask the manufacturer to give the median (or fiftieth percentile) of his or her distribution. In other words, we ask for that value of θ which would lead him or her to be equally willing to bet on the proportion being above or below this value.

(c) Having obtained the median, which divides his or her distribution in half, we ask the manufacturer to divide it further into quarters, by specifying the twenty–fifth and seventy–fifth percentiles. To do this for the seventy–fifth percentile, for example, we ask the manufacturer to concentrate on those values of θ lying between the median and the upper limit, and then to choose a value in this range such that the probabilities of θ lying above or below this value are assessed to approximately equal.

At the end of this interrogation process, we have five points on the cumulative distribution function of θ; namely, the first, twenty-fifth, fiftieth, seventy-fifth and ninety-ninth percentiles. A smooth curve can be drawn through these to get a reasonable approximation to the manufacturer's subjective cumulative distribution function (c.d.f).

Figure 7 shows such a function corresponding to the following (hypothetical) elicited values:

(a) 0.05, 0.6; (b) 0.2; (c) 0.125, 0.3.

The corresponding histogram for subintervals 0.05–0.1, 0.1–0.15, etc. is shown in Figure 8 together with a smoothed approximation to the probability density function (p.d.f).

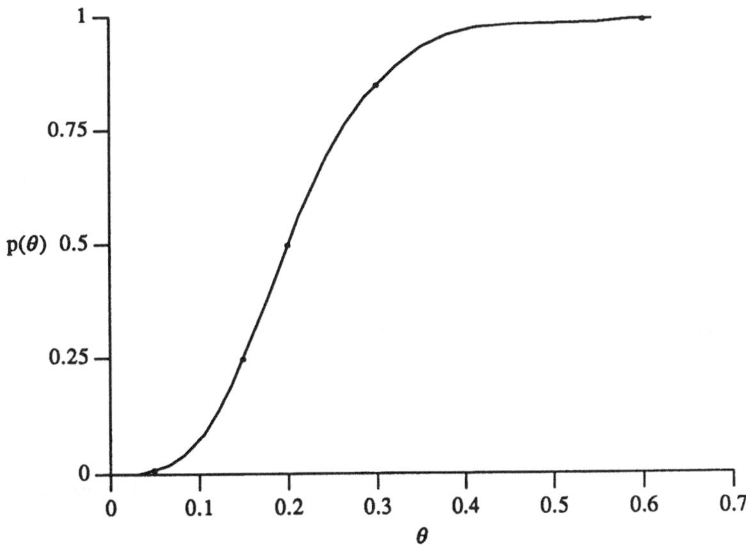

Figure 7. Hypothetical subjective c.d.f..

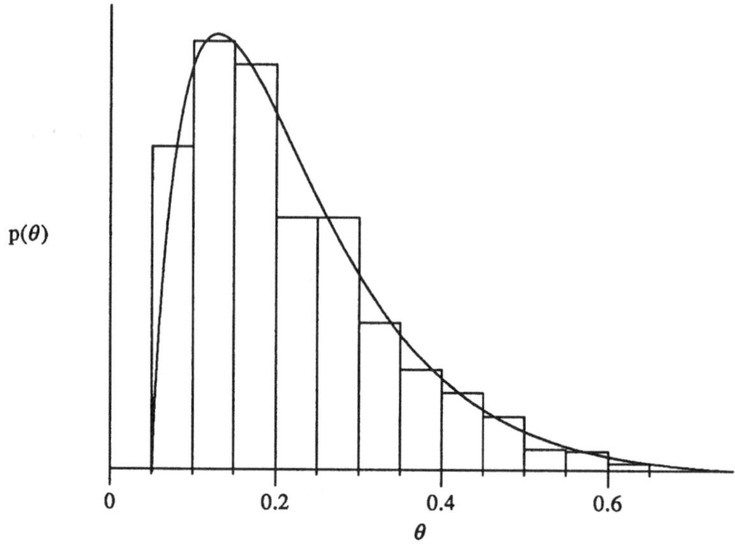

Figure 8. Histogram and smoothed p.d.f. corresponding to Figure 7.

We have been considering the assessment of a prior distribution for a single parameter. The assessment of a joint distribution for two or more parameters is straightforward if, *a priori*, assessment can be considered independently (since we then reduce to a series of one-dimensional assessments), but can present considerable difficulties if there are complex dependencies among the parameters.

15. Bayesian Inference for some Univariate Probability Models

Inferences for the Binomial and related distributions

If y denotes the number of 'successes' obtained in n independent 'trials', each with chance θ of a 'success', then if n is fixed in advance we obtain the Binomial probability model

$$f(y|\theta,n) = \binom{n}{y} \theta^y(1-\theta)^{n-y}, \qquad y = 0, 1, ..., n$$

Let us suppose that prior beliefs about θ ($0 \le \theta \le 1$) are specified in the form of a density $p(\theta)$. The posterior density is then given by

$$p(\theta|y,n) = \frac{f(y|\theta,n)p(\theta)}{\int_0^1 f(y|\theta,n)p(\theta)\, d\theta} \qquad (0 \le \theta \le 1).$$

The form of $p(\theta|y,n)$ can easily be computed numerically *for any choice of* $p(\theta)$. However, this is tiresome if we wish to explore how $p(\theta|y,n)$ varies for *different choices of* $p(\theta)$, since we require a separate numerical excercise for each specification. Moreover, simply producing a curve numerically fails to provide any analytic insight into the manner in which the data and prior beliefs interact to form posterior beliefs.

For these reasons, it is of interest to examine a particular *mathematical* representation of $p(\theta)$ (whilst bearing in mind the practical approaches to assessing *actual* forms of $p(\theta)$ discussed earlier). In order to obtain mathematical representations which afford both theoretical and practical insights, it would be very convenient if we could discover a *family of probability densities* which, by varying the small number of parameters of a mathematical function, could be made to generate a range of 'shapes' of prior beliefs which would adequately represent many actual forms of belief occurring in practice.

In the Binomial case, we require a family of functions defined on the interval $0 \le \theta \le 1$, and specified in terms of just a few parameters, which can be varied to provide a number of flexible forms. Such a family of densities is the *beta family*, defined, for $a>0$, $b>0$, by

$$p(\theta) = \frac{\Gamma(a+b)}{\Gamma(a)\Gamma(b)}\, \theta^{a-1}(1-\theta)^{b-1} \qquad (0 \le \theta \le 1),$$

where, $\Gamma(.)$ is the gamma function, having the property $\Gamma(z+1) = z\Gamma(z)$, $z>0$. Since $p(\theta)$ is a density, we have $\int p(\theta)d\theta = 1$, and so,

$$\int_0^1 \theta^{a-1}(1-\theta)^{b-1}d\theta = \frac{\Gamma(a)\Gamma(b)}{\Gamma(a+b)}$$

By varying a and b, a very wide family of shapes can be generated. With the advent of interactive computing, it is possible to interrogate a subject in order to discover whether (and, if so, for which a,b) his or her beliefs about

θ can be adequately represented by a beta density.

For a particular choice of a,b, the proportional form of Bayes' Theorem gives

$$p(\theta|y,n) \propto f(y|\theta,n)p(\theta)$$

$$\propto \theta^{y}(1-\theta)^{n-y}.\ \theta^{a-1}(1-\theta)^{b-1}$$

$$\propto \theta^{a+y-1}(1-\theta)^{b+n-y-1}$$

Hence, we have

$$p(\theta|y,n) = \frac{\theta^{a+y-1}(1-\theta)^{b+n-y-1}}{\int_0^1 \theta^{a+y-1}(1-\theta)^{b+n-y-1}\, d\theta}$$

$$= \frac{\Gamma(a+b+n)}{\Gamma(a+y)\Gamma(b+n-y)}\ \theta^{a+y-1}(1-\theta)^{b+n-y-1},$$

using the result noted above in order to evaluate the integral.

Now we have an interesting result. *The posterior density is also of the beta form*, but with parameters a+y and b+n−y in place of the prior parameters a,b. Schematically, we may write this as

---

$\frac{\text{Prior}}{\text{Beta (a,b)}}$ with $\frac{\text{Likelihood}}{\text{Binomial(n,y)}}$ implies $\frac{\text{Posterior}}{\text{Beta(a+y,b+n−y)}}$

This result provides a very simple rule for updating beliefs in the case of a Binomial probability model and prior beliefs represented by a beta density. The above analysis then suffices as a once–and–for–all solution for any choice of a,b and for any data n,y. Provision of credible intervals, for example, can straightforwardly be made by reference to suitable tables of the beta distribution, or by noting that the quantity

$$\left(\frac{b+n-y}{a+y}\right)\cdot\left(\frac{\theta}{1-\theta}\right)$$

has an $F_{2(a+y),2(b+n-y)}$ distribution, so that the tables of the F-distribution may be utilized.

Some insight into the way in which posterior inferences combine prior information with that contained in the data is obtained by examining the form of the point estimate of θ provided by the mean of the posterior distribution. The posterior mean is given by

$$\hat{\theta} = \int_0^1 \theta p(\theta|y,n)\, d\theta$$

$$= \frac{\Gamma(a+b+n)}{\Gamma(a+y)\Gamma(b+n-y)} \int_0^1 \theta^{(a+y+1)-1}(1-\theta)^{(b+n-y)-1} d\theta$$

$$= \frac{\Gamma(a+b+n)}{\Gamma(a+y)\Gamma(b+n-y)} \cdot \frac{\Gamma(a+y+1)\Gamma(b+n-y)}{\Gamma(a+b+n+1)}$$

$$= \frac{a+y}{a+b+n}$$

using results stated above for the integral and for the gamma function.

When we examine $\hat{\theta}$ more closely, we see that it can be rewritten in the form

$$\hat{\theta} = \frac{(a+b)\left(\frac{a}{a+b}\right) + n\left(\frac{y}{n}\right)}{a+b+n}$$

$$= (1-w)\left(\frac{a}{a+b}\right) + w\left(\frac{y}{n}\right),$$

where $w = n/(a+b+n)$.

This reveals that the mean of the posterior distribution is a *weighted average* of two quantities. $a/(a+b)$ and y/n. The former is, in fact, *the mean of the prior distribution* (which can be shown by direct calculations, or can be deduced from the form of the posterior mean with $n = 0$, $y = 0$); the latter is the *'natural' estimate of θ using the data alone* (and is also the maximum likelihood estimate and MVUE).

The posterior estimate thus combines what the data tells us, y/n, with our 'best guess' before seeing the data, $a/(a+b)$. As the amount of data increases, that is as n becomes larger and larger, the weight w which attaches to the data–estimate, y/n, becomes larger:

$$\frac{n}{a+b+n} = \frac{1}{1+(a+b)/n} \to 1 \text{ as } n \to \infty$$

Conversely, if we have no data, that is $n = 0$, we use the prior estimate $a/(a+b)$ (since $w = 0$). The form of the posterior estimate therefore adapts itself, in an intuitively sensible way, taking account of the amount of data available for modifying prior beliefs.

In addition to the summary of the posterior density provided by a point estimate, we could also look at what happens to the 'spread' of the posterior as n increases. If we assess the spread by looking at the variance of the posterior distribution, we shall require the form of the variance of a beta distribution with parameters $(a+y, b+n-y)$. Using standard methods for finding variances, this is easily shown to be equal to

$$[(a+y)(b+n-y)]/[(a+b+n)^2(a+b+n+1)].$$

As $n \to \infty$, this variance clearly tends to 0. Thus, whatever the particular prior choice of a and b, as the amount of data increases, beliefs become more and more concentrated around the posterior mean. The latter itself comes more and more to resemble y/n (as we have just seen) and so, eventually, no matter which particular beta prior individuals adopt, they would all come to believe more and more strongly that the true chance of success θ is 'very close' to the observed frequency of successes y/n.

More specifically, if y and $n - y$ become very large compared with a and b, the posterior density will be (approximately) beta with parameters $(y, n-y)$. Thus, the effect of a large amount of data will have been to force a range of different prior beliefs (represented by many different choices of a and b) into a posterior *consensus*. This is the key to the Bayesian answer to accusations of the lack of 'objectivity' in Bayesian methods as a result of the 'subjective' intrusion of prior beliefs. For the Bayesian, it is natural to regard subjective beliefs as primary, and 'objective consensus' as a special case, arising when the amount of data available is large enough to overwhelm and dominate everyone's prior beliefs, forcing them into the same posterior shape. Recall the illustration given in Section 12.

In cases where the amount of data is *not* large enough to force this kind of consensus, a Bayesian statistician regards it as right and proper that prior beliefs will have an effect on posterior inferences. If conditions for achieving 'consensus' are not satisfied, there is no merit in pretending that there is an 'objective' answer. All that a statistician can do in such cases is to display a *range* of posterior inferences corresponding to a *range* of different kinds of prior belief. The reader (of a scientific report) or client (of a statistical consultant) can then assess his or her own reaction to the data by identifying the particular prior to posterior analyses that correspond most closely to his or her own prior beliefs. Such a display of analyses is greatly facilitated (as we have just seen) by exploiting a *mathematical* (rather than a purely numerical) approach using a flexible, tractable family of prior distributions. If a family can be found which is (a) rich enough to represent most 'shapes' of prior belief occurring in practice, and (b) 'fits nicely' with the likelihood, so that the mathematical form of the posterior density is easily identified, the inference process is then completely summarized by noting how the parameters of the prior density transform to those of the posterior. In the case of the beta density, we saw that the prior parameters (a,b) are transformed into posterior parameters (a+y, b+n−y), the data entering into this transformation via the sufficient statistics, n and y.

Suppose now that having observed y successes in n trials, we are interested in 'predicting' the number of successes that will result in a further m independent trials. If $\tilde{x}$ denotes this future number of successes, a Bayesian statistician will wish to calculate

$$\pi(x|m,y,n) = P(\tilde{x} = x|m,y,n), \qquad x = 0,1,\ldots,m.$$

Using straightforward probability calculus, these probabilities are given by

$$\pi(x|m,y,n) = \int_0^1 f(x|m,\theta)p(\theta|y,n)d\theta,$$

an example of a *predictive* distribution.

In the case we are considering, assuming a prior beta (a,b) distribution for θ, we obtain, for $x = 0,1,\ldots,m$,

$$\pi(x|m,y,n) = \binom{m}{x}\frac{\Gamma(a+b+n)}{\Gamma(a+y)\Gamma(b+n-y)} \cdot \int_0^1 \theta^x(1-\theta)^{m-x}\theta^{a+y-1}(1-\theta)^{b+n-y-1}d\theta$$

$$= \binom{m}{x}\frac{\Gamma(a+b+n)}{\Gamma(a+y)\Gamma(b+n-y)} \cdot \int_0^1 \theta^{a+y+x-1}(1-\theta)^{b+n-y+m-x-1}d\theta$$

$$= \binom{m}{x}\frac{\Gamma(a+b+n)}{\Gamma(a+y)\Gamma(b+n-y)} \cdot \frac{\Gamma(a+y+x)\Gamma(b+n-y+m-x)}{\Gamma(a+b+n+m)}$$

Example. If $a = b = 1$, so that $p(\theta)$ is taken to be a uniform density over the interval (0,1), and if $m = x = 1$, so that we are considering the probability that a single further trial will turn out to be a success, $\pi(x|m,y,n)$ simplifies to give

$$P(\tilde{x} = 1|m = 1,y,n) = \frac{\Gamma(n+2)}{\Gamma(y+1)\Gamma(n-y+1)} \cdot \frac{\Gamma(y+2)\Gamma(n-y+1)}{\Gamma(n+3)}$$

$$= \frac{y+1}{n+2}$$

using the fact that $\Gamma(z+1) = z\Gamma(z)$.

The results discussed in this chapter have been developed on the basis of the Binomial probability model. However, models such as the Negative–Binomial will lead to identical forms of $p(\theta|y,n)$, for any choice of $p(\theta)$.

Inferences for the Poisson Distribution

If $x = (x_1, x_2, ..., x_n)$ is a random sample from a Poisson distribution with parameter θ, we have the probability model

$$f(x|\theta) = \prod_{i=1}^{n} \frac{\theta^{x_i} e^{-\theta}}{x_i!} \qquad (x_i \geq 0,\ i=1,2,...,n)$$

$$= \frac{\theta^{n\bar{x}}\ e^{-n\theta}}{\prod (x_i!)}$$

where $n\bar{x} = (x_1 + x_2 + ...+ x_n)$.

To carry out the Bayesian analysis, we must specify a prior probability density $p(\theta)$ for θ, the parameter appearing in the Poisson distribution. In this case, θ can be any positive real number and so

$$p(\theta|x) = \frac{f(x|\theta)p(\theta)}{\int_0^\infty f(x|\theta)p(\theta)d\theta}, \qquad 0 \le \theta < \infty.$$

As we argued in the previous subsection, it would be very convenient if we could find a family of probability density functions, generating a wide range of shapes of possible prior beliefs as we vary the parameters of the family, and 'fitting together' in a tractable way with the likelihood defined by the Poisson model.

Such a family is provided by the *gamma densities*, which have the form

$$p(\theta) = \frac{b^a \theta^{a-1} e^{-b\theta}}{\Gamma(a)} \qquad (0 \le \theta < \infty)$$

for any choice of $a > 0$, $b > 0$. As we vary a and b, a wide variety of shapes can be generated. Since $p(\theta)$ is a density, we have $\int_0^\infty p(\theta)d\theta = 1$, and hence

$$\int_0^\infty \theta^{a-1} e^{-b\theta} d\theta = \Gamma(a)/b^a$$

Using this result, it is easy to show that the mean of the gamma distribution is equal to a/b, and the variance is equal to a/b^2. By appropriately choosing the parameters a,b, a shape can be chosen to reflect the location and spread of a wide range of actual prior beliefs.

For a particular choice of a,b, the proportional form of Bayes' Theorem gives

$$p(\theta|x) \propto f(x|\theta)p(\theta)$$

$$\propto \theta^{n\bar{x}} e^{-n\theta}\theta^{a-1}e^{-b\theta}$$

$$\propto \theta^{a+n\bar{x}-1} e^{-(b+n)\theta}$$

Hence, we have

$$p(\theta|x) = \frac{\theta^{a+n\bar{x}-1} \; e^{-(b+n)\theta}}{\int_0^\infty \theta^{a+n\bar{x}-1} \; e^{-(b+n)\theta} \; d\theta}$$

$$= \frac{(b+n)^{a+n\bar{x}} \; \theta^{a+n\bar{x}-1} \; e^{-(b+n)\theta}}{\Gamma(a+n\bar{x})}$$

the expression for the integral being obtained from the form noted above.

Comparing the forms of $p(\theta)$ and $p(\theta|x)$, we note that the latter is also a gamma density, and that we have demonstrated a general result which can be expressed schematically in the form

Prior		Likelihood		Posterior
Gamma(a,b)	with	Poisson $(n\bar{x})$	implies	Gamma $(a+n\bar{x}, b+n)$

This result provides a very simple rule for updating beliefs in the case of a Binomial probability model and prior beliefs represented by a gamma density. The parameters of the latter are simply transformed using the sufficient statistics n and $\bar{x}$.

Posterior credible intervals can easily be derived using tables of the χ^2 distribution by noting that the quantity $2(b+n)\theta$ has a χ^2 distribution with $2(a+n\bar{x})$ degrees of freedom, a fact which is easily demonstrated by the standard transformation technique

The posterior mean, a possible choice of point estimate, is given by

$$\hat{\theta} = \int_0^\infty \theta p(\theta|x) d\theta$$

$$= \frac{(b+n)^{a+n\bar{x}}}{\Gamma(a+n\bar{x})} \int_0^\infty \theta^{(a+n\bar{x}+1)-1} \; e^{-(b+n)\theta} \; d\theta$$

$$= \frac{(b+n)^{a+n\bar{x}}}{\Gamma(a+n\bar{x})} \cdot \frac{\Gamma(a+n\bar{x}+1)}{(b+n)^{a+n\bar{x}+1}}$$

$$= \frac{a+n\bar{x}}{b+n}$$

using the fact that $\Gamma(z+1) = z\Gamma(z)$.

Noting that we can write

$$\frac{a+n\bar{x}}{b+n} = \frac{b}{b+n} \cdot \frac{a}{b} + \frac{n}{b+n} \cdot \bar{x}$$

we see again that the posterior mean is a weighted average of the *prior estimate* of θ (a/b) and the *data-based estimate* ($\bar{x}$). As n becomes large, the weight attached to $\bar{x}$ becomes larger and approaches 1. Moreover, the posterior variance $(a+n\bar{x})/(b+n)^2$ tends to 0 and so beliefs become more and more concentrated around $\bar{x}$, irrespective of the precise initial choice of a,b, provided the latter are small compared with $n\bar{x}$ and n. In this case, a possibly widely different set of prior beliefs will be forced to a posterior consensus of beliefs, well represented by a gamma distribution with parameters $n\bar{x}$ and n. The mean and variance of this posterior distribution are given by $\bar{x}$ and $\bar{x}/n$, respectively.

If we are interested in a predictive distribution for the next observation to be made, $y = x_{n+1}$, we calculate

$$\pi(y|x) = \int_0^\infty f(y|\theta)p(\theta|x)d\theta$$

$$= \frac{(b+n)^{a+n\bar{x}}}{y!\Gamma(a+n\bar{x})} \cdot \int_0^\infty \theta^{a+n\bar{x}+y-1} e^{-(b+n+1)\theta} d\theta$$

$$= \frac{(b+n)^{a+n\bar{x}}}{y!\Gamma(a+n\bar{x})} \cdot \frac{\Gamma(a+n\bar{x}+y)}{(b+n+1)^{a+n\bar{x}+y}}$$

Example. Suppose y = 0, so that we are interested in the probability of the next observation being zero, then

$$P(\tilde{y} = 0|x) = \left[\frac{b+n}{b+n+1} \right]^{a+n\bar{x}}$$

This is seen to be an intuitively sensible form. All other things being equal, small values of $n\bar{x}$ (i.e. a small total number of observations on the n previous occasions) will lead to values close to 1 (particularly if n is large). On the other hand, if $n\bar{x}$ is large the ratio (b+n)/(b+n+1) (which is less than 1) is raised to a high power and a small probability is obtained.

16. Approximate Analysis under Great Prior Uncertainty

Considering, for convenience of exposition, a single unknown parameter θ, situations in which great prior uncertainty exists (relative to the information contained in the data) correspond, mathematically, to a prior density function which is relatively flat in the region where the (standardized) likelihood, corresponding to data x, is peaked. This may arise from a moderate sized experiment in conjunction with very diffuse prior beliefs, or even from a very large experiment with moderately strong prior beliefs.

Recalling that Bayes' theorem may be summarized in the form:

"posterior density = standardized likelihood X prior density",

we see that if the prior density is approximately constant as a function of θ, over the range where the likelihood is concentrated, then, in this case of relative great prior uncertainty, Bayes' theorem gives the approximate result

$$p(\theta|x) \approx \frac{f(x|\theta)}{\int f(x|\theta)d\theta} \propto f(x|\theta)$$

since $p(\theta) \approx$ constant appears in both numerator and denominator of Bayes' theorem and so cancels.

We see, therefore, that, when prior beliefs are relatively weak, posterior beliefs are dictated by the location and shape of the likelihood. In particular, we see that the value, $\hat{\theta}$, of θ which is considered "most likely", given the data x, is the value which maximizes the likelihood $f(x|\theta)$: that is, the *maximum likelihood estimate*, much used by non-Bayesian statisticians (cf. Newby, chapter 3, section 2.2).

The approximate argument given thus far tells us that, in the case of great prior uncertainty, posterior beliefs will peak around the maximum likelihood estimate, $\hat{\theta}$, of θ. In fact, if we are able to accept a few more approximating assumptions, we can not only learn about the *location*, but also about the *shape* of the likelihood, and thus of the approximate posterior density.

To see this, let us recall from the argument given above that, approximately,

$$p(\theta|x) \propto exp\{L(\theta)\}$$

where

$$L(\theta) = \log\{f(x|\theta)\}$$

Now let us suppose that $L(\theta)$ can be well-approximated by considering a Taylor series expansion about $\hat{\theta}$, the series expansion being continued as far as the quadratic term. In other words, we are assuming that the logarithm of the likelihood function can be well approximated by a quadratic function in the vicinity of the maximum likelihood estimate, $\hat{\theta}$. Assuming this form of approximation, we have

$$L(\theta) \approx L(\hat{\theta}) + (\theta-\hat{\theta})\, L'(\hat{\theta}) + ½\, (\theta-\hat{\theta})^2\, L''(\hat{\theta}),$$

where $L'(\hat{\theta})$, $L''(\hat{\theta})$ denote the first and second derivatives of $L(\theta)$, with respect to θ, evaluated at $\theta = \hat{\theta}$. If we further assume that $f(x|\theta)$ has a unique maximum at $\hat{\theta}$, then so does $L(\theta)$, since taking the logarithm of a function does not change the location of its turning points; in particular, it follows that $L'(\hat{\theta}) = 0$ (the derivative at the maximum is zero).

Noting that, as a function of θ, $L(\hat{\theta})$ is constant, we have the approximation

$$L(\theta) = \text{constant} - \frac{(\theta-\hat{\theta})^2}{2(-L''(\hat{\theta}))^{-1}}$$

The motivation for rewriting the quadratic term in this way becomes clear if we note that

$$p(\theta|x) \propto exp\left\{-\frac{1}{2\hat{\sigma}^2}(\theta-\hat{\theta})^2\right\}$$

where $\hat{\sigma}^2 = (-L''(\hat{\theta}))^{-1}$.

It follows from the form of $p(\theta|x)$ that, provided the assumptions we have made are not unreasonable, ***posterior beliefs about θ are well approximated by a normal density with mean $\hat{\theta}$ and variance $\hat{\sigma}^2$***. The *location* of posterior beliefs is therefore determined by $\hat{\theta}$, the maximum likelihood estimate, and the *spread* of posterior beliefs is inversely proportional to minus the second derivative of the log-likelihood at $\hat{\theta}$. This latter is really quite an intuitive measure of spread: the second derivative is measuring how quickly the gradient of the log-likelihood is changing(from positive to negative, hence the minus sign): if the gradient changes quickly, this indicates that the likelihood is sharply peaked and thus the spread is small.

Example: *Binomial likelihood.* It can easily be checked that, in this case, approximate posterior beliefs about θ are represented by an $N\left\{\frac{x}{n}, \frac{1}{n}\left[\frac{x}{n}\left(1-\frac{x}{n}\right)\right]\right\}$ density, provided the above assumptions are reasonable.

Example: *Poisson likelihood.* If x is the total number of occurrences from n independent observations of a Poisson process with parameters θ, posterior beliefs are approximated by an $N\left\{\frac{x}{n}, \frac{1}{n}\left[\frac{x}{n}\right]\right\}$ density.

17. Problems Involving Many Parameters: Empirical Bayes

When models contain *many* parameters, even a seemingly "large" sample may, in effect, not really contain overwhelming evidence about the unknown aspects of the model, since the information is "spread" over many parameters. On the other hand, when a model contains many parameters it will typically be the case that we have substantial information about the ***relationships*** that exist among such parameters. After all, parameters usually "represent" something "real", and the individual "real" elements are likely to be highly interconnected or we should not be considering them altogether in one model in the first place!

In such situations, specification of independent, vague priors to each parameter would usually *not* be a genuine representation of prior information. And yet typical non–Bayesian procedures correspond to Bayesian procedures derived using just such a prior specification. It follows that in many–parameter situations there is a great deal to scope for finding Bayesian forms of inferences that will differ considerably from standard forms, as the following simple examples show.

Suppose that data consist of k groups of observations, n in each group, and that all observations may be assumed normally and independently distributed with equal variances (known or unknown; it does not matter) and unknown group means θ_1, θ_2, ..., θ_k. Given these assumptions, how should the parameters be estimated?

Standard procedures (for example, least squares or maximum likelihood) would lead to the use of the sample means $\bar{x}_1$, $\bar{x}_2$, ..., $\bar{x}_k$, as estimates of θ_1, θ_2, ..., θ_k. This would also be the solution from a Bayesian point of view if a joint prior specification for $(\theta_1, \theta_2, ..., \theta_k)$ had the following form:

(a) $p(\theta_1,\theta_2,...,\theta_k) = \prod_i p(\theta_i)$, so that beliefs about any individual θ_i are *independent* of beliefs about other θ_j's; this means that all the θ_j's are regarded as *unrelated* parameters;

(b) each $p(\theta_i)$ corresponds to a *vague prior specification.*

But is this a *realistic* form of prior specification? Consideration of some concrete examples falling into the general structure of k groups with n observations in each suggests not.

For example, suppose the observations are logarithms of failure times of components, the groups corresponding to *slightly* different manufacturing conditions. Suppose further that there are twenty groups (k = 20) with two components in each group (n = 2). In this case, it seems intuitively much more sensible to consider estimating θ_i by an estimate of the form $w\bar{x}_i + (1-w)\bar{x}$, where $0 < w < 1$ and $\bar{x}$ is the overall mean based on data from all the groups. This is a *weighted average* of information drawn from just the ith group $(\bar{x}_i)$ and overall information $(\bar{x})$, and reflects a feeling that n = 2 is small compared with kn = 40 and that it might be inefficient to use only information from two observations if we really believed that the effects of the different manufacturing conditions should only be *slightly* different. The weight w should reflect the relative sizes of k and n, as well as some measure of how "similar" we feel the groups to be.

As a second example, suppose, instead, that observations within each group were revealed to be replicate, observed log failure time responses to a given level of test temperature, the temperature being fixed for each group, but increasing as we proceed from the first to the kth group (with, say, levels $s_1 < s_2 < \dots < s_k$).

In such contexts, the temperature levels may span the range over which it is known that responses tend to decrease, flatten out, and finally increase (i.e. failures tend to occur more quickly at very low or very high temperatures). If we again take $n = 2$, $k = 20$, many people would be unhappy, intuitively, with using $\bar{x}_i$ to estimate the true response θ_i corresponding to temperature s_i. Some might suggest fitting, say, a quadratic curve through the plot of the $\bar{x}_i$ against s_i, and then using the fitted value corresponding to s_i as an estimate. Others might favour a weighted average between this fitted value and the group mean $\bar{x}_i$, the relative weights depending in the general case on k and n.

The detail of these particular examples is not really the issue here. The point to note is that background details of the real situation provide information very different from that encapsulated in the "independent, vague prior" form discussed above. The Bayesian conclusion is that knowledge of the relationships that hold between parameters by virture of their *meanings* should be incorporated into the model through a prior distribution. Mathematically, this suggests using a *hierarchical* model of the form

$$\begin{cases} p(x|\theta_1,\theta_2,\dots\theta_k) \\ p(\theta_1,\theta_2,\dots,\theta_k|\phi) \\ p(\phi) \end{cases}$$

where the first stage relates parameters to observations, the second specifies the nature of the relationship existing among the parameters, and the third stage incorporates numerical information (if any) about the general form of relationship specified in the second stage.

As an example, the temperature/failure situation may be modelled by considering

$$\begin{cases} \log x_{ij} \sim N(\theta_i, \sigma^2) & i = 1,\dots,k, \; j = 1,\dots,n \\ \theta_i \sim N(\phi_0+\phi_1 s_i+\phi_2 s_i^2, \tau^2) & i = 1,\dots,k, \\ p(\phi_0,\phi_1,\phi_2) \approx \text{constant.} & \end{cases}$$

The second stage here would represent the information that true failure time response means lie approximately on a quadratic curve, while the third stage would indicate vagueness about the precise numerical form of the quadratic.

It can be shown that the posterior mean for θ_i from such a model is of the form

$$w\bar{x}_i + (1-w)(\hat{\phi}_0+\hat{\phi}_1 s_i+\hat{\phi}_2 s_i^2)$$

where $w = n\tau^2/(n\tau^2+\sigma^2)$ and $\hat{\phi}_0 + \hat{\phi}_1 s + \hat{\phi}_2 s^2$ denotes the quadratic curve fitted by least squares through the means $\bar{x}_i$.

Generally, if we examine the above hierarchical model more carefully, in many applications the detailed form of the first two stages is as follows.

$$\begin{cases} p(x|\theta_1,\dots,\theta_k) = \prod_i p(x_i|\theta_i) \\ p(\theta_1,\dots,\theta_k|\phi) = \prod_i p(\theta_i|\phi) \end{cases}$$

This arises when we have data from k "similar" contexts, with $p(x_i|\theta_i)$ providing the probability model for the ith context and the θ_i regarded as a "random sample from a population distribution".

Straightforward use of probability calculus now reveals that, given ϕ,

$$p(\theta_i|\text{data},\phi) = p(\theta_i|x_i,\phi) \propto p(x_i|\theta_i)\; p(\theta_i|\phi),$$

whereas

$$p(\theta_i|\text{data}) = \int p(\theta_i|x_i,\phi)\; p(\phi|\text{data})d\phi$$

with

$$p(\phi|\text{data}) \propto \Pi \int p(x_i|\theta_i)\, p(\theta_i|\phi)d\theta_i \cdot p(\phi)$$

Overall uncertainty about individual context parameters, θ_i, is thus seen to involve the whole data, via $p(\phi|\text{data})$, rather than just the directly related data, x_i. A "pooling of information" across the different contexts (for example, units or subsystems) thus takes place.

If $p(\phi|\text{data})$ is sharply peaked around its mode, $\hat{\phi}$, say, we have the approximate result

$$p(\theta_i|\text{data}) \approx p(\theta_i|x_i,\hat{\phi})$$

Thus, it is *as if* we took the form $p(\theta_i|x,\phi)$, which would obtain if ϕ were known, and substituted into it an estimate of ϕ based on all the data. In effect, it is as if we "first used the data to estimate the prior", giving $p(\theta_i|\hat{\phi})$, and then applied Bayes' Theorem.

This form of approximation to the full Bayesian analysis of the hierarchical model is often referred to as *Empirical Bayes* analysis. Viewed in the above light, this is not a new form of statistical methodology, but rather an approximate Bayesian method.

The key idea here is that confronted with data from many different, but "similar", sources, we have a formal model and methodology for pooling the information in order to achieve two things: first, to learn something, via $p(\phi|\text{data})$, about the average, overall behaviour of the "population" (of units, subsystems, or whatever) under study; secondly, to "borrow strength" from the whole set of data when learning about the characteristics, $p(\theta_i|\text{data})$, of individual "members of the population".

Further illustration of such hierarchical/Empirical Bayes analysis is given in the next section.

18. Numerical Methods for Practical Bayesian Statistics

Given a likelihood $f(x|\theta)$ and prior density $p(\theta)$, where x and θ (both typically vector-valued) denote data and unknown parameters, respectively, the starting point for inferences about θ is the joint posterior density for θ

given by

$$p(\theta|x) = \frac{f(x|\theta)\ p(\theta)}{\int f(x|\theta)p(\theta)d\theta}.$$

In fact, of course, we are usually interested in summaries of the full joint posterior distribution. For example, attention may be focussed on univariate marginal densities for some or all of the components θ_i of θ; bivariate joint marginal densities for various pairs (θ_i,θ_j) of component parameters; or even on simpler summaries in the form of posterior first and second moments. Alternatively, we may be interested in posterior summaries for functions of one or more of the components of θ: for example, marginal and joint densities for θ_i/θ_j and $\theta_i\theta_j$.

In all these cases, the technical key to the implementation of the formal solution given by Bayes' theorem, for specified likelihood and prior, is the ability to perform a number of integrations. First, we need to evaluate the denominator in order to obtain the normalising constant of the posterior density; then we need to integrate over complementary components of θ, or transformations of θ, in order to obtain marginal (univariate or bivariate) densities, together with summary moments, highest posterior density intervals and regions, or whatever. Except in certain rather simple problems the required integrations will not be feasible analytically and so efficient numerical strategies will be required. Finally, the finite sets of numerical values obtained after marginalisation need to be reconstructed into a graphical representation of a univariate or a bivariate marginal posterior distribution.

An iterative quadrature strategy

It is well known that univariate integrals of the type

$$\int_{-\infty}^{\infty} e^{-t^2} f(t)dt$$

are often well-approximated by Gauss-Hermite quadrature rules of the form

$$\sum_{i=1}^{n} w_i f(t_i)$$

where t_i is the ith zero of the Hermite polynomial $H_n(t)$. In particular, if $f(t)$ is a polynomial of degree at most 2n–1, then the approximation is without error. It follows that, if $h(t)$ is a suitably well-behaved function and

$$g(t) = h(t)(2\pi\sigma^2)^{-\frac{1}{2}} exp\left\{- ½ \left[\frac{t-\mu}{\sigma}\right]^2\right\}$$

then

$$\int_{-\infty}^{\infty} g(t)dt \approx \sum_{i=1}^{n} m_i g(z_i),$$

where

$$m_i = w_i \; exp\{t_i^2\} \sqrt{2}\sigma, \quad z_i = \mu + \sqrt{2}\sigma t_i$$

(see Naylor & Smith, 1982). We see, therefore, that, expressed in informal terms, Gauss–Hermite rules are likely to prove very efficient for functions which closely resemble 'polynomial X normal' forms. In fact, this is a rather rich class which, even for moderate n (≤11, say), covers many of the likelihood X prior shapes we typically encounter for parameters defined on $(-\infty, \infty)$. Moreover, the applicability of this approximation can be greatly extended by working with suitable transformations of parameters defined on other ranges, such as $(0,\infty)$ or (a,b), using, for example, $\log(t)$ or $\log(t-a)-\log(b-t)$, respectively. Of course, to use the formulae we must specify μ and σ. It turns out that, given reasonable starting values (from any convenient source; prior information, maximum likelihood estimates etc.), we can successfully iterate, substituting estimates of the posterior mean and variance obtained based on previous values of m_i and z_i. Moreover, we note that if the posterior density is well–approximated by the product of a normal and a polynomial of degree at most 2n–3, then an n–point Gauss–Hermite rule will prove effective for *simultaneously evaluating the normalising constant and the first and the second moments*, using the same (iterated) set of m_i and z_i. In practice, it is efficient to begin with a small grid size (n=4 or n=5) and then to gradually increase the grid size until stable answers are obtained both within and between the last two grid sizes used.

Our discussion so far has been for the one dimensional case. Clearly, however, the need for an efficient strategy is most acute in higher dimensions. The 'obvious' extension of the above ideas is to use a cartesian product rule, giving the approximation

$$\int \dots \int f(t_1, \dots, t_k) dt_1 \dots dt_k \simeq \sum_{ik} m_{ik}^{(k)} \dots \sum_{i1} m_{i1}^{(1)} g(z_{i1}^{(1)}, \dots, z_{ik}^{(k)}),$$

where the grid points and weights,

$z_{ij}^{(j)}$ and $m_{ij}^{(j)}$, respectively,

are found by substituting the iterated estimates of μ and σ^2 corresponding to the marginal component t_j.

The problem with this 'obvious' strategy is that the product form is only efficient if we are able to make an (at least approximate) assumption of posterior independence among the individual components. In such cases, the lattice of integration points formed from the product of the one-dimensional grids is efficiently covering the bulk of the posterior density. However, this would not be the case if there were high posterior correlation, since this would lead to many of the lattice points falling in areas of negligible posterior density, thus causing poor estimates of the normalising constant and of moments.

To overcome this problem, we can apply individual parameter transformations of the type discussed above, and then attempt to transform the resulting parameters, via an appropriate linear transformation, to a new approximately orthogonal, set of parameters. At the first step, this linear transformation derives from an initial guess or estimate of the posterior covariance matrix (for example, based on the observed information matrix from a maximum likelihood analysis). Successive transformations are then based on the estimated covariance matrix from the previous iteration.

We are led to the following general strategy:

(1) Reparametrise individual parameters so that the resulting working parameters all take values on the real line.
(2) Using initial estimates of the joint posterior mean vector and covariance matrix for the working parameters, transform further to a centred, scaled,

more 'orthogonal' set of parameters.

(3) Using the derived initial location and scale estimates for these 'orthogonal' parameters, carry out, on suitably dimensioned grids, cartesian product integration of functions of interest.
(4) Iterate, successively updating the mean and covariance estimates, until stable results are obtained both within and between grids of specified dimension.

For full details and illustrations of the method, see Naylor & Smith (1982) and Smith *et al.* (1987).

A sampling–resampling perspective

In the continuous case, the integration operation plays a fundamental role in Bayesian statistics; be it for calculating the normalizing constants, marginal distributions or moments. Moreover, realistic choices of likelihood and prior will necessitate the use of sophisticated numerical integration techniques of the kind just described.

An alternative way of addressing this problem is to take a new look at Bayes theorem from a sampling–resampling perspective. This will be seen to open the way both to easily implemented calculations and to essentially calculus–free insight into the mechanics and uses of Bayes theorem.

As a first step, we note the essential duality between a sample and the density (distribution) from which it is generated. Clearly, the density generates the sample; conversely, given a sample we can approximately recreate the density (as a histogram, a kernel density estimate, an empirical c.d.f. or whatever).

Suppose we now shift the focus in Bayes' theorem from densities to samples. In terms of densities, the inference process is encapsulated in the updating of the prior density, $p(\theta)$, to the posterior density, $p(\theta|x)$, through the medium of the likelihood function, $f(x|\theta)$. Shifting to samples, this corresponds to the updating of a sample from $p(\theta)$ to a sample from $p(\theta|x)$ through the likelihood function $f(x|\theta)$. As a preliminary to discussing how we might do this, consider the following general problem. Suppose that a sample of random variates is easily generated, or has already been generated, from a continuous density $g(\theta)$, but that what is really required is a sample from a density $h(\theta)$. Can we somehow utilize the sample from $g(\theta)$ to form a sample from $h(\theta)$? Slightly more generally, given a positive function $f(\theta)$ which is normalizable to such a density, $h(\theta) = f(\theta)/\int f(\theta)d\theta$, can we form a sample from the latter, given only a sample from $g(\theta)$ and the functional form of $f(\theta)$?

We can approximately resample from $h(\theta) = f(\theta)/\int f(\theta)d\theta$ as follows. Given θ_i, $i = 1,...,n$, a sample from g, calculate $\omega_i = f(\theta_i)/g(\theta_i)$ and then $q_i = \omega_i/(w_1+...+w_n)$. Draw θ^* from the discrete distribution over $\{\theta_1,...,\theta_n\}$ placing mass q_i on θ_i. Then θ^* is approximately distributed according to h with the approximation 'improving' as n increases. This procedure is a variant of the by now familiar bootstrap resampling procedure. The usual bootstrap provides equally likely resampling of the θ_i, while here we have weighted resampling with weights determined by the ratio of f to g, again in agreement with intuition. See Rubin (1988).

How does Bayes' theorem generate a posterior sample from a prior sample? For fixed x, define $f_x(\theta) = f(x|\theta)p(\theta)$. Then with $g(\theta) = p(\theta)$, we may immediately apply the weighted bootstrap resampling method to obtain samples from the density corresponding to the standardized f_x, which is precisely the posterior density $p(\theta|x)$. Thus, we see that Bayes' theorem, as a mechanism for generating a posterior sample from the prior sample, takes the following simple form: for each θ_i in the prior sample accept θ_i into the posterior sample with probability

$$q_i = f(x|\theta_i)/\sum_{j=1}^{n} f(x|\theta_j)$$

otherwise reject it.

The likelihood therefore acts as a resampling probability; those θ in the prior sample having high likelihood are more likely to be retained in the posterior sample.

Several obvious uses of this sampling-resampling perspective are immediate. Using large prior samples and iterating the resampling process for successive individual data elements illustrates the sequential Bayesian learning process, as well as the increasing concentration of the posterior as the amount of the data increases. In addition, the approach provides natural links with elementary graphical displays; e.g. histograms, stem and leaf displays, boxplots to summarize univariate marginal posterior distributions, scatterplots to summarize bivariate posteriors, etc.

An important issue in Bayesian inference is sensitivity of inferences to model specification. In particular we might ask:

how does the posterior change if we change the prior?
how does the posterior change if we change the likelihood?

In the density function / numerical integration setting, such sensitivity studies are rather off-putting, in that each change of a functional input typically requires one to carry out new calculations from scratch. This is not the case with the sampling-resampling approach, as we now illustrate in relation to the questions posed above.

In comparing two models in relation to the second question, we note that changes in likelihood may arise in terms of

(i) change in distributional specification with θ retaining the same interpretation, e.g. a location,
(ii) change in data to a larger data set (prediction), a smaller data set (diagnostics), or a different data set (validation).

To unify notation, we shall in either case denote two likelihoods by $f_1(x|\theta)$ and $f_2(x|\theta)$. We denote two different priors to be compared in relation to the first question by $p_1(\theta)$ and $p_2(\theta)$. For complete generality, we shall consider changes to both likelihood and prior, although in any particular application we would not typically change both. Denoting the corresponding posterior densities by $p_1(\theta|x)$, $p_2(\theta|x)$ we easily see that

$$p_2(\theta|x) \propto \frac{f_2(x|\theta)\ p_2(\theta)}{f_1(x|\theta)\ p_1(\theta)}\ p_1(\theta|x)$$

Letting $v_x(\theta)$ denote the ratio on the right-hand side, we note that to implement the weighted bootstrap method we can take $g(\theta) = p_1(\theta|x)$, $f_x(\theta) = v_x(\theta)p_1(\theta|x)$ and $\omega_i \propto v_x(\theta_i)$. Resampled θ^* will then be approximately distributed according to the standardized $f_x(\theta)$, which is precisely $p_2(\theta|x)$.

Again, different aspects of the sensitivity of the posteriors to changes in inputs are easily studied by graphical examination of the posterior samples.

The Gibbs sampler

The Gibbs sampler is a Markovian updating scheme enabling one to obtain (in the limit) samples from a joint distribution, via iterated sampling from full conditional distributions. Given a joint posterior density $p(\theta|x)$, functional forms of the k univariate full conditional densities (i.e. the distribution of each individual component of θ conditional on specified values of the data x and all the other components) can be readily written down, at least up to a proportionality. If, suppressing for notational convenience, the dependence on x, these full conditional densities are denoted by

$$p(\theta_1|\theta_2,\theta_3,\ldots,\theta_k)$$
$$p(\theta_2|\theta_1,\theta_3,\ldots,\theta_k)$$
$$\ldots\ldots \quad \ldots\ldots$$
$$p(\theta_k|\theta_1,\theta_2,\ldots,\theta_{k-1})$$

then the Gibbs sampling algorithm proceeds as follows. Choose initial values $\theta_2^{(0)},\theta_3^{(0)},\ldots,\theta_k^{(0)}$ and generate a value $\theta_1^{(1)}$ from the conditional density

$$p(\theta_1|\theta_2^{(0)},\theta_3^{(0)},\ldots,\theta_k^{(0)})$$

Similarly, generate a value $\theta_2^{(1)}$ from the conditional density

$$p(\theta_2|\theta_1^{(1)},\theta_3^{(0)},\ldots,\theta_k^{(0)})$$

and continue up to the value $\theta_k^{(1)}$ from the conditional density.

$$p(\theta_k|\theta_1^{(1)},\theta_2^{(1)},\ldots,\theta_{k-1}^{(1)})$$

Then, with the new realisation of θ, $\theta^{(1)}$, replacing the initial values the above process is iterated, say m times, producing $\theta^{(m)}$. Under mild mathematical conditions on the form of the problem, for large m $\theta_i^{(m)}$ can be regarded (approximately) as a simulated realisation from $p(\theta_i)$, the marginal distribution of θ_i. Independent parallel replication of the entire above process t times, produces t sets of parameter values, $(\theta_j^{(m)} \equiv \theta_j,\ j = 1,\ldots,t)$ and thus for each element of θ we obtain a simulated sample of size t from its marginal density. Graphical reconstruction of marginal densities can proceed using either a kernel density estimate or by averaging over the conditional density:

$$p(\theta_i) \approx \frac{1}{t}\sum_{j=1}^{t} p\left(\theta_i | \theta_j^{[i]}\right)$$

where $\theta^{[i]}$ denotes the complement of θ_i, in θ. See Gelfand & Smith (1990).

Suppose interest centres on the marginal distribution for ϕ which is a (typically nonlinear) function $g(\theta_1,\ldots,\theta_k)$ of $\theta_1,\ldots,\theta_k$. We note that evaluation of g at each of the sampled sets of θ's provides samples of ϕ, so that a kernel density estimate can readily be calculated.

Complete implementation of the Gibbs sampler requires that a determination of t be made and that, across iterations, choice(s) of m be specified. In this regard it is important to distinguish the assessment of convergence for any individual data application from the broader goal of developing on-line, automated, interactive software to determine satisfactory convergence.

Extensive experience with a wide range of particular applications suggests that accomplishing the former is not a problem. We note that appropriate values for t and m will depend upon the particular application and cannot be specified in advance. Since random variate generation is generally inexpensive we expect to experiment with different settings.

The generated data is monitored in a univariate fashion allowing the sampler to run until the marginal posterior distributions for each parameter of interest "converge". This is done in an elementary manner. For a fixed m we increase t, overlay plots of the resulting estimated densities, and see if the estimates are visually indistinguishable. Similarly, we also increase m to assess stability of the density estimate.

Consider a general Bayesian hierarchical model having 3 stages. In an obvious notation, we write the joint distribution of the data and parameters as

$$p(y|\theta)\ p(\theta|\phi)\ p(\phi)$$

where we assume all components of prior specification to be available for sampling. Primary interest is usually in the marginal posterior $p(\theta|y)$.

As a concrete illustration, consider an hierarchical Poisson model. Suppose we observe independent counts, s_i, over differing lengths of time t_i (with resultant rate $\rho_i = s_i/t_i$), $i = 1,\ldots,p$. Assume $p(s_i|\lambda_i)$ = Poisson $(\lambda_i t_i)$ and that λ_i are i.i.d. from Gamma(α,β), with density $\lambda_i^{\alpha-1}e^{-\lambda_i/\beta}/\beta^{\alpha}\Gamma(\alpha)$. The parameter α will be assumed known (in practice, we might treat α as a 'tuning' parameter or perhaps, in an empirical Bayes spirit, estimate it from the marginal distribution of the s_i's) and β, in turn, is assumed to arise from an Inverse Gamma distribution IG$((\gamma,\delta)$ with density $\delta^{\gamma}e^{-\delta/\beta}/\beta^{\gamma+1}\Gamma(\gamma)$. (A diffuse version of this final stage distribution is obtained by taking δ and γ to be very small, perhaps zero.)

Letting $y = (s_1,\ldots,s_p)$, the conditional distributions $p(\lambda_j|y)$ are sought. The posterior of λ_j conditional on the data and all other parameters is given by

$$p(\lambda_j|y,\beta,\lambda_{i,i\neq j}) = G\left[\alpha + s_j, \left(t_j + \frac{1}{\beta}\right)^{-1}\right] \qquad j = 1,\ldots,p$$

while the posterior of β conditional on the data and all other parameters is given by

$$p(\beta|y,\lambda_1,\ldots,\lambda_p) = IG(\gamma+p\alpha,\Sigma\lambda_i+\delta).$$

Given $(\lambda_1^{(0)},\lambda_2^{(0)},\ldots,\lambda_p^{(0)},\ \beta^{(0)})$ the Gibbs sampler draws

$$\lambda_j^{(1)} \sim G\left[\alpha + s_j, \left(t_j + \frac{1}{\beta^{(0)}}\right)^{-1}\right],\ j = 1,\ \ldots,\ p,$$

and then

$$\beta^{(1)} \sim \mathrm{IG}\left[\gamma + \alpha p, \sum_{j=1}^{p} \lambda_j^{(1)} + \delta\right]$$

to complete one cycle. If we carry out m repetitions each of i iterations, generating $(\lambda_{1l}^{(i)}, \ldots, \lambda_{pl}^{(i)}, \ldots., \beta_l^{(i)})$, $l = 1, \ldots, m$, the marginal density estimate for λ_j is given by

$$\frac{1}{m}\sum_{l=1}^{m} G\left[\alpha + s_j, \left(t_j + \frac{1}{\beta_l^{(i)}}\right)^{-1}\right] \qquad j = 1, \ldots, p$$

while that for β is given by

$$\frac{1}{m}\sum_{l=1}^{m} \mathrm{IG}(\gamma+\alpha p, \sum \lambda_{jl}^{(i)} + \delta)$$

In the general hierarchical model notation, $\theta_1 = (\lambda_1, \ldots, \lambda_p)$, $\phi = \beta$.

Example. An application of this model and methodology to a real life reliability problem is reported in Gelfand & Smith (1990), section 4.2. The problem considered involves failure data on 10 similar pumps. Individual pump failure data is modelled by Poisson distributions. The underlying failure rates are assumed to be sampled from a Gamma (α,β) distribution with α known. Finally, β is assigned an inverse gamma prior distribution. In the resulting analysis, data from all the pumps is effectively "pooled", in a coherent manner, to improve inferences about the failure rates of individual pumps.

References

DeGroot, M.H. (1970), *Optimal Statistical Decisions*, McGraw–Hill.

Gelfand, A.E. & Smith, A.F.M. (1990), Sampling based approaches to calculating marginal densities, *Journal of the American Statistical Association*, **85**, 398–409.

Lindley, D.V. (1985), *Making Decisions (Second Edition)*, Wiley, New York.

NAYLOR, J.C. & SMITH, A.F.M. (1982), Applications of a method for the efficient computation of posterior distributions,. *Applied Statistics*, **31**, 214–225.

RUBIN, D.B. (1988). Using the SIR algorithm to simulate posterior distributions. In: J.M. BERNARDO *et al.* (Eds), *Bayesian Statistics* 3, Oxford University Press.

SMITH, A.F.M., SKENE, J.E.H. & NAYLOR, J.C. (1987), Progress with numerical and graphical methods for practical Bayesian statistics, *Statistician*, **36**, 75–82.

DEPARTMENT OF MATHEMATICS
HUXLEY BUILDING
IMPERIAL COLLEGE
180 QUEEN'S GATE
LONDON
UK SW7 2BZ

3. RELIABILITY MODELLING AND ESTIMATION

by

MARTIN NEWBY

Eindhoven University of Technology

Abstract

Some simple models for the analysis of data from non-repairable systems are described and applied in some reliability and failure rate estimation problems. The analyses are used to illustrate the Bayesian approach to estimation. The importance of the link between the likelihood, prior and posterior is emphasised and examples are given to show the influence of the prior and the likelihood on the estimators obtained from the posterior density.

1. Non-Repairable Systems

1.1 Introduction

A non-repairable system is typified by a simple component such as a light bulb. The item is put into service and ends its service when it fails. Thus the non-repairable system is often an appropriate model for analysis at the component level. Other systems may be analysed as non-repairable when their mode of operation allows it. For instance, military equipment may be required to operate only for the duration of a mission, after the mission it is extensively maintained. In such a case the equipment may be regarded as new at the beginning of each cycle of operation.

In analysing the behaviour of non-repairable systems data from a variety of sources has to be processed. Data may come from laboratory experiments, field trials, field data, maintenance records and so on. In many cases only partial information is available because experiments are terminated before all items have failed, or because records of field data are incomplete. Often time and cost constraints mean that tests are terminated prematurely and again analysis must be based on incomplete data.

P. Sander and R. Badoux (eds.), Bayesian Methods in Reliability, 81–100.

1.2. Describing reliability

For a non–repairable system the following measures of reliability are used.

(i) *Hazard rate* a measure of aging. The hazard rate gives the conditional probability of failing in the next instant of time.

(ii) *Mean residual life* is the mean life left for a working system. It is useful for models which have costs specified.

(iii) The *reliability* is the probability of surviving longer than a specified time under a given set of conditions.

The hazard rate and mean residual life are the most helpful characterisations of component behaviour, for an increasing hazard rate, or decreasing mean residual life, imply wearout or aging. Conversely, a decreasing hazard rate or increasing mean residual life indicate an improvement over time.

The measures of reliability are all equivalent as is shown below.

Define (cf. Smith section 4):

hazard rate	$h(t)$
reliability	$R(t)$
mean residual life	$\mu(t)$
probability density	$f(t)$
cumulative distribution	$F(t)$
cumulative hazard	$H(t)$

then specifying any one uniquely specifies the others.

$F(t) + R(t) = 1$ because they are complementary probabilities

$$F(t) = \int_0^t f(u)du; \quad f(t) = F'(t) = \frac{d}{dt} F(t)$$

$$h(t) = \frac{f(t)}{R(t)} = -\frac{d}{dt} \{\ln R(t)\}$$

$$H(t) = \int_0^t h(u)du$$

$$R(t) = exp\ \{-H(t)\}$$

$$\mu(t) = \frac{1}{R(t)} \int_t^{\infty} R(u)du$$

$$R(t) = \frac{\mu(0)}{\mu(t)} \, exp \, \{- \int_0^t \frac{1}{\mu(x)} \, dx\}$$

$$h(t) = \frac{\mu'(t) + 1}{\mu(t)}$$

These theoretical relations are often difficult to use in practise, some cannot be obtained in closed form and some are not easily estimated from data.

Mostly we are concerned with estimation and inference for failure time distributions based on samples of observed failure times.

1.3. Failure time distributions

A small number of failure time distributions is commonly used in reliability. They are used partly because they are mathematically tractable and partly because they correspond to intuitive ideas about failure mechanisms. The distributions have been described in more detail by Smith (cf. chapter 2 section 4), they are listed below.

The *exponential* $R(t) = exp \, \{-\lambda t\}$ $\lambda > 0, \, t \geq 0$

The *gamma* $f(t) = \frac{\lambda(\lambda t)^{k-1}}{\Gamma(k)} \, exp \, \{-\lambda t\}$ $\lambda > 0, \, t \geq 0$

The *Weibull* $R(t) = exp \, \{-(t/b)^k \}$ $k, \, b > 0, \, t \geq 0$

The *log-normal* $R(t) = 1 - \Phi\left(\frac{\ln(t) - \mu}{\sigma}\right)$

$$= 1 - \Phi\left(\frac{\ln(t/b)}{\sigma}\right) \qquad \mu = \ln(b), \, b > 0, \, \sigma > 0, \, t \geq 0$$

and Φ is the standard normal distribution.

The *inverse-Gaussian.* This is also a version of the Wald distribution.

$$R(t) = \Phi[(\lambda/t)^{\frac{1}{2}} \, (1 - t/\mu)] - e^{2\lambda/\mu} \, \Phi[-(\lambda/t)^{\frac{1}{2}} \, (1 + t/\mu)]$$

$$\mu, \, \lambda > 0, \, t \geq 0$$

2. Estimation

2.1. Introduction

When data is collected on failure or survival a list of times is obtained. Some of the times are failure times and others are the times at which the subject left the experiment. These times both give information about the performance of the system. The two types will be referred to as *failure* and *censoring* times (cf. Smith section 5).

A censoring time, t^*, gives less information than a failure time, for it is known only that the item survived past t^* and not when it failed. The data is thus collected as a list of failure times t_1, t_2, ..., t_n and of censoring times t_1^*, t_2^*, ..., t_m^*.

2.2. Classical methods

The failure times are assumed to follow a parametric distribution $F(t;\theta)$ with density $f(t;\theta)$ and reliability $R(t;\theta)$. There are several methods of estimating the parameter θ based only on the data in the sample without any prior assumptions about θ. The availability of powerful computers and software packages has made the method of maximum likelihood the most popular. Descriptions of most methods can be found in the book by Mann, Schafer and Singpurwalla (1974). In general the method of maximum likelihood is the most useful of the classical approaches.

The likelihood approach is based on constructing the joint probability distribution or density for a sample. Thus for the sample with failure times t_1 , t_2, ..., t_n, censoring times t_1^*, t_2^*, ..., t_m^*, and assuming independence, the likelihood is a function

$$L(\text{sample};\theta) = \prod_{i=1}^{n} f(t_i;\theta) \cdot \prod_{i=1}^{m} R(t_i^*;\theta)$$

The product of the densities is the joint density of n failure times. The reliabilities appear in the second product because it is known only that failure occurs after t_i^* with $R(t_i^*;\theta)$ as the probability of survival past t_i^*. The likelihood is thus proportional to the probability of obtaining the given sample from the assumed distribution.

In the classical view, the likelihood is regarded as a function of the parameter θ, and estimates of θ are obtained by choosing values which give the highest probability of getting the sample. In other words an estimate $\hat{\theta}$ of θ is the value of θ which maximises L(sample;θ) as a function of θ. The estimate $\hat{\theta}$ of θ is chosen so that

$$L(\text{sample};\hat{\theta}) \geq L(\text{sample};\theta)$$

for all θ.

Maximum likelihood is useful because in most cases $\hat{\theta}$ is asymptotically normally distributed. A second useful property is that of invariance. Thus if $\omega = \phi(\theta)$ is a one to one change of parameter, then the maximum likelihood estimation of ω, $\hat{\omega}$, is given by

$$\hat{\omega} = \phi(\hat{\theta})$$

In particular if we are interested in the hazard rate, or reliability, then the maximum likelihood estimators of $\hat{h}(t;\theta)$ of $h(t;\theta)$ and $\hat{R}(t;\theta)$ of $R(t;\theta)$ are

$$\hat{h}(t;\theta) = h(t;\hat{\theta})$$
$$\hat{R}(t;\theta) = R(t;\hat{\theta})$$

While these properties are useful, the asymptotic normality may require rather large samples before the estimator is approximately normally distributed and the asymptotic variance–covariance matrix may be hard or impossible to obtain. Thus probability statements and confidence regions are hard to construct.

By contrast, the Bayesian approach regards the likelihood as a conditional probability, the probability of seeing the sample given the value of θ. Thus we write the likelihood as

$$L(\text{sample}|\theta).$$

2.3. Bayesian methods

The parameter θ is now assumed to be a random variable with density function $g(\theta)$. Because of this the failure time density is written $f(t|\theta)$, a density conditional on θ. In what follows some examples of Bayesian methods in life testing will be given.

Example 1

Suppose that the failure density is $f(t|\theta)$ and that θ is a random variable with density $g(\theta)$. One interpretation is that the observed failure time density is just the expected density with respect to θ. When $f(t|\theta)$ is exponential

$$f(t|\theta) = \theta \, exp\{-\theta t\}$$

and θ has a gamma density

$$g(\theta) = \frac{\theta^{\alpha-1} \, exp\{-\theta/\beta\}}{\beta^{\alpha} \, \Gamma(\alpha)}$$

the observed failure density is

$$f_1(t) = E_g \, [\theta e^{\theta t}] = \int_0^{\infty} \theta e^{-\theta t} \cdot \frac{\theta^{\alpha-1} \, e^{-\theta/\beta}}{\beta^{\alpha} \, \Gamma(\alpha)} \, d\theta$$

$$= \frac{\alpha\beta}{(\beta t + 1)^{\alpha+1}}$$

The observed reliability is

$$R_1(t) = E \, [e^{-\theta t}] = \int_0^{\infty} e^{-\theta t} \cdot \frac{\theta^{\alpha-1} \, e^{-\theta/\beta}}{\beta^{\alpha} \, \Gamma(\alpha)} \, d\theta$$

$$= \frac{1}{(\beta t + 1)^{\alpha}}$$

The observed times thus have a Pareto distribution. This model is closely related to that used by Littlewood & Verrall (1973) for software reliability (cf. Littlewood section 3.4).

In general we face the problem of assessing the models through estimates of statistics obtained by Bayesian methods. Now the classical likelihood is regarded as the joint probability of the observed data conditional on the parameter and is written $L(data|\theta)$. The central tool is the relationship

$$f(\theta|sample) = k\ L(sample|\theta)g(\theta) \tag{1}$$

where f is the posterior density, g the prior density and L the likelihood. The constant k,

$$k^{-1} = \int_{\theta} L(sample|\theta)g(\theta)d\theta$$

is a normalising constant.

In the following most conclusions derive from (1). The posterior density is of most interest here because the reliability examples require the estimation of parameters of distributions or point estimates of reliability itself.

Throughout the examples estimators will be assumed to be based on a squared error loss function. The squared error loss function gives the mean of the posterior density as the appropriate point estimate. If we chose the absolute error as the loss function then the median would be the appropriate estimate (see Martz & Waller(1982)).

3. Reliability Estimation

3.1. Introduction

From elementary reliability considerations, the reliability function, R(t), is the probability of operating (surviving) for longer than time t. If we fix the time period t there is probability $p=R(t)$ of success and $q=1-p=1-R(T)=F(t)$ of failure. In this case the reliability at time t is then simply a binomial probability model with probability p of success of failures. The model is built up as follows.

- n: number of items on test, fixed in advance
- t: the test period, fixed in advance
- x: the observed number of failures, a random variable.

If the reliability over time t is denoted by p, then the binomial probability of x failures is

$$f(x|p) = \binom{n}{x} (1 - p)^x \, p^{n-x} \qquad 0 \le x < n, \quad 0 \le p \le 1$$

The problem is to estimate the reliability, p, from the observed number of failures or survivors.

The simplicity of the model is gained at the expense of losing the actual failure times. For example if a component has a Weibull reliability function with mean life 100 hrs and shape parameter 3 the probability of surviving at least 120 hrs is

$$p = exp\left\{-\left(\frac{120}{112}\right)^3\right\} = 0.21$$

and the probability of failure within 120 hrs is thus 0.79 = 1 − 0.21. If a sample of size 10 is selected and put on test, then the distribution of the number of failures in the first 120 hrs is the binomial

$$b[x;10,0.21] = \binom{10}{x} (0.89)^x \, (0.21)^{n-x}$$

Turning this model round and making the reliability the object of interest makes binomial sampling a means of reliability estimation.

3.2. Binomial sampling

The objective is now to estimate the reliability of an item over a fixed period of time. This is accomplished by setting the sample on test for a fixed time period and observing the number of failures. The object of interest is now the reliability, p = R(t). The model is as follows.

n: the number of items on test, fixed in advance

t: the test period, fixed in advance

x: the number of survivors at time t, a random variable.

Notice that x has been defined as the number of survivors, the model can equally well use the number of failures, n−x, as the variable to observe.

Since the reliability at time t is p, the probability of x survivors is

$$f(x|p) = \binom{n}{x} p^x (1-p)^{n-x}, \qquad 0 \le x \le n, \quad 0 \le p \le 1$$

The classical approach takes the maximum likelihood estimate $\hat{p} = x/n$ as the estimate of the reliability.

However, if information about p is available in the form of a prior probability density, g(p), the Bayesian approach may be used. The density f(x|p) is also the likelihood, so the posterior density becomes

$$g(p|x) = \frac{f(x|p)g(p)}{f(x)}$$

$$= \frac{p^x (1-p)^{n-x} g(p)}{\int_0^1 p^x (1-p)^{n-x} g(p)\, dp}$$

Example 2

For simplicity, consider a uniform prior

$$g(p) = 1 \qquad 0 \le p \le 1$$
$$g(p) = 0 \qquad \text{elsewhere}$$

This prior is an unlikely choice since it represents a state of ignorance, however, the argument is straightforward to follow.

$$g(p|x) = \frac{p^x (1-p)^{n-x}}{\int_0^1 p^x (1-p)^{n-x}\, dp}$$

$$= \frac{p^x(1-p)^{n-x}}{\beta(x+1,n-x+1)}$$

where $\beta(.,.)$ is the beta function.

Thus g(p|x) is a beta density (see Martz & Waller(1982)) with parameters x+1 and n−x+1. The squared error loss function gives the expected value of p as the estimator, so the Bayes estimator here is

$$\tilde{p} = E(p|x) = \int_0^1 p\ g(p|x)dp = \frac{x+1}{n+2}$$

Notice that the estimate is only a little different from the maximum likelihood estimator $\hat{p} = x/n$, which is, in fact, the mode of the posterior density.

Suppose that in 10 trials there are 6 survivors. The two estimates of the reliability, p, are

Bayes: $\tilde{p} = \frac{7}{12} = 0.58$

maximum likelihood: $\hat{p} = \frac{6}{10} = 0.60$

Example 3

A more appropriate choice of prior for p is a beta density. It is appropriate because, firstly it can give more weight to one value of p and secondly it is a conjugate prior for this problem. Conjugate priors are easier to use because the posterior and the prior density have the same form.

Take the prior as

$$g(p;\ell,m) = \frac{1}{\beta(\ell,m)} p^{\ell-1}(1-p)^{m-1} \qquad \ell,m > 0, \quad 0 \le p \le 1$$

The mean of the prior is

$$\bar{p} = \frac{\ell}{\ell+m}$$

and the mode is

$$p_m = \frac{\ell-1}{\ell+m-2}$$

so that knowledge about the most likely values of p can be represented by suitable choices of ℓ and m. Beta densities for a few values of ℓ and m are shown in Figure 1.

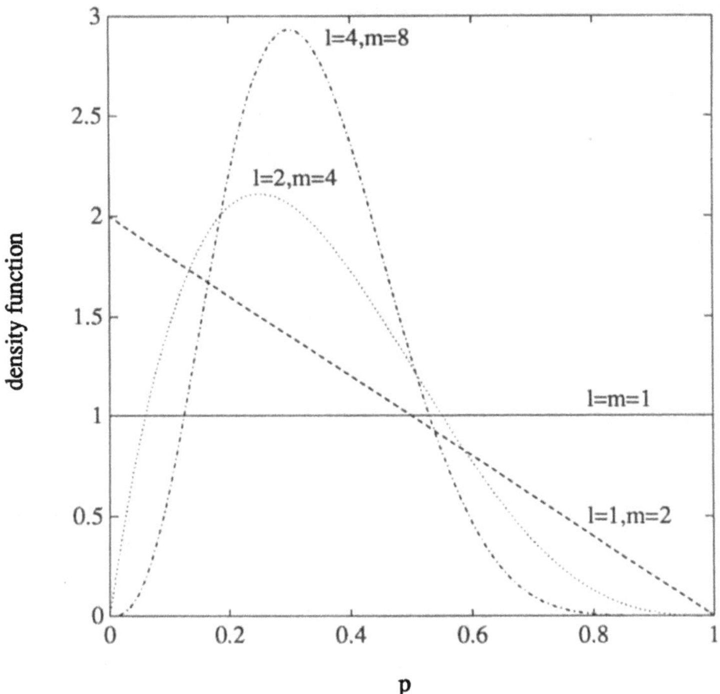

Figure 1. Beta densities.

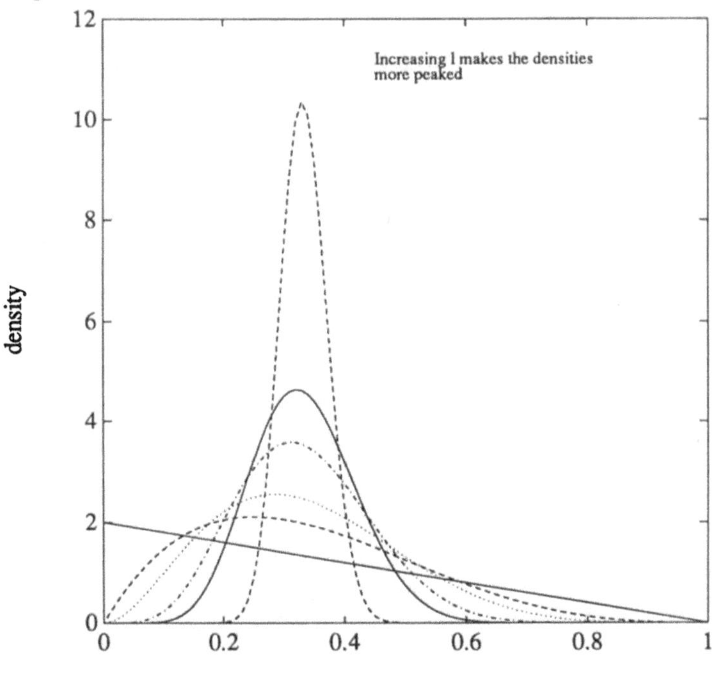

Figure 2. Beta density with mean 1/3.

In this example the posterior density is the beta

$$g(p|x) = \frac{1}{\beta(\ell+x, m+n-x)} p^{\ell+x-1}(1-p)^{m+n-x-1}$$

The point estimator of p is the posterior mean

$$\tilde{p} = E(p|x) = \frac{\ell+x}{\ell+m+n}$$

To continue the example use the previous data, that is 6 survivors out of 10 observed in a life test of fixed duration. The posterior mean is then

$$\tilde{p} = \frac{\ell+6}{\ell+m+10}$$

This example can be used to show the effect of the strength of prior belief on the posterior result. Suppose that the prior mean of p is $\frac{1}{3}$ but that there are varying degrees of certainty. Values of ℓ and m can be chosen to reflect this situation. If ℓ is given, then taking $m = 2\ell$ ensures that the prior mean is $\frac{1}{3}$. The greater the value of ℓ the more concentrated is the density around the mean. The beta densities with mean $\frac{1}{3}$ are shown in Figure 2. The densities become more sharply peaked as ℓ increases.

Using ℓ to define beta densities with mean $\frac{1}{3}$ gives a posterior estimate

$$\tilde{p} = \frac{\ell+6}{3\ell+10}$$

And this can be used to show the effect of the prior on the posterior. Recall that the maximum likelihood estimate is $\hat{p} = 0.6$ which is shown in Figure 3, increasing ℓ indicates increasing certainty in the prior. The figure shows how the increasing concentration in the prior pulls the posterior estimate towards the prior mean. Clearly in the limit the effects of the data are lost and the prior dominates.

3.3. Pascal sampling

In binomial sampling the sample size and the test duration were fixed in advance. Clearly this has implications for the cost of the experiment. In some circumstances it is appropriate to set the duration of the test and the number of failures (or survivors) to be observed. Now the sample size, n, becomes a random variable. The procedure is to test the items, one at a time, for a

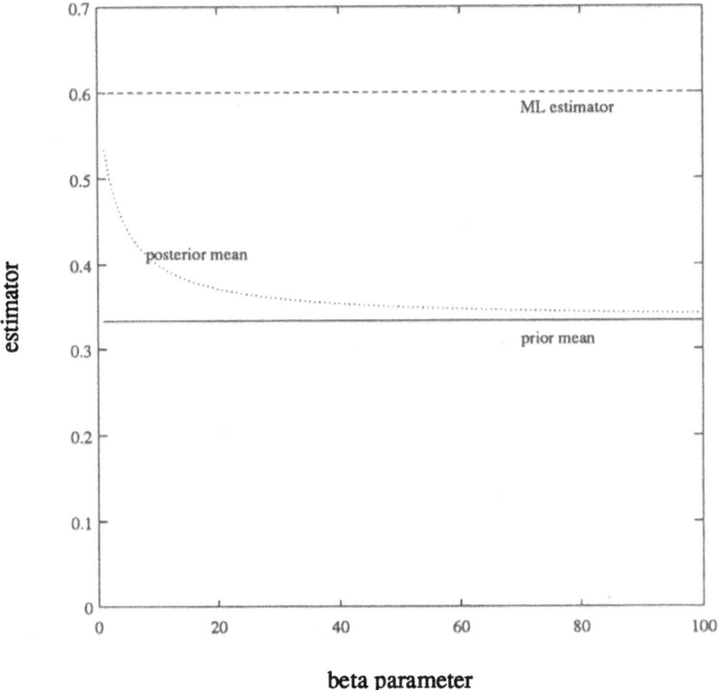

Figure 3. Interaction of prior and likelihood.

period t. The failure or survival of the item is noted. The model then deals with

t: the test period, fixed in advance

x: the number that survive the test period, fixed in advance

n: the total number of items tested, a random variable.

The situation is thus a waiting time problem. The distribution of n is obtained by noting that there must be exactly x–1 successes recorded in n–1 trials and that the n–th trial yields a success. Thus the probability function for n based on survivors is

$$f(n|p) = \left\{ \binom{n-1}{x-1} p^{x-1} (1-p)^{(n-1)-x-1} \right\} p$$

$$= \binom{n-1}{x-1} p^{x} (1-p)^{n-x}$$

As before the maximum likelihood estimate of p is $\hat{p} = \frac{x}{n}$.

Beta prior

Using the prior

$$g(p) = \frac{1}{\beta(\ell,m)} p^{\ell-1}(1-p)^{m-1}$$

gives the posterior density

$$g(p|n) = \frac{1}{\beta(\ell+x,m+n-x)} p^{\ell+x-1}(1-p)^{m+n-x-1}$$

another beta density, formally the same as the case for binomial sampling. The posterior mean, used to estimate p, is

$$E(p|x) = \frac{\ell+x}{\ell+m+n}$$

Again, the estimate of p is formally the same as for binomial sampling, but the natures of x and n have changed, x is a fixed number and n the random variable.

Example 4

A design of valve spring is to be tested in order to determine its reliability over 10^8 operation cycles. It is decided that 10 survivors need to be observed. It is found that after 11 trials 2 springs have failed and 9 springs survived. At the twelfth trial the spring survives. Now the 10 survivors have been observed and 12 trials were required. Thus x=10 and n=12. The density for n is

$$f(n|p) = \binom{11}{9} p^{10}(1-p)^2$$

where p is the reliability over 10 cycles.

A discussion with the users of earlier designs gives a strong indication that the reliability over 10^8 cycles should be close to 0.8. Thus a prior for p is chosen with mean 0.8 and a small variance. The chosen prior is

$$g(p) = \frac{1}{\beta(48,12)} p^{47}(1-p)^{11}$$

Notice that if the indications had been less certain a prior with mean 0.8 but a greater variance is (cf. Smith section 16)

$$g_1(p) = \frac{1}{\beta(4,1)} p^3(1-p)^0 = 4p^3$$

Using the chosen prior, the posterior density is

$$g(p|n) = \frac{1}{\beta(58,14)} p^{57}(1-p)^{13}$$

And the estimate of p is

$$\tilde{p} = E(p|n) = \frac{58}{72} = 0.81$$

The maximum likelihood estimate is

$$\hat{p} = \frac{10}{12} = 0.83$$

and had the less certain $\beta(4,1)$ density been used the estimate would have been

$$\tilde{p} = \frac{14}{17} = 0.82$$

And once more the interaction of the data and the prior may be seen. The less certain prior gives an estimate closer to the likelihood estimate, and the more certain prior pulls it towards the prior mean. The sample likelihood and the densities are sketched in Figure 4.

3.4. Poisson sampling

The previous examples gave no help in discovering how failures were distributed in time. All that was known was the duration of the test and that the item failed or survived the test. However, it might be necessary to know how the failures were distributed in time, were they uniformely spread or did they all come late in the interval? If a specific distribution is assumed for the survival times then answers to these questions can be sought. The simplest model is obtained by assuming that the failure times follow the exponential density

$$f(t|\lambda) = \lambda e^{-\lambda t}$$

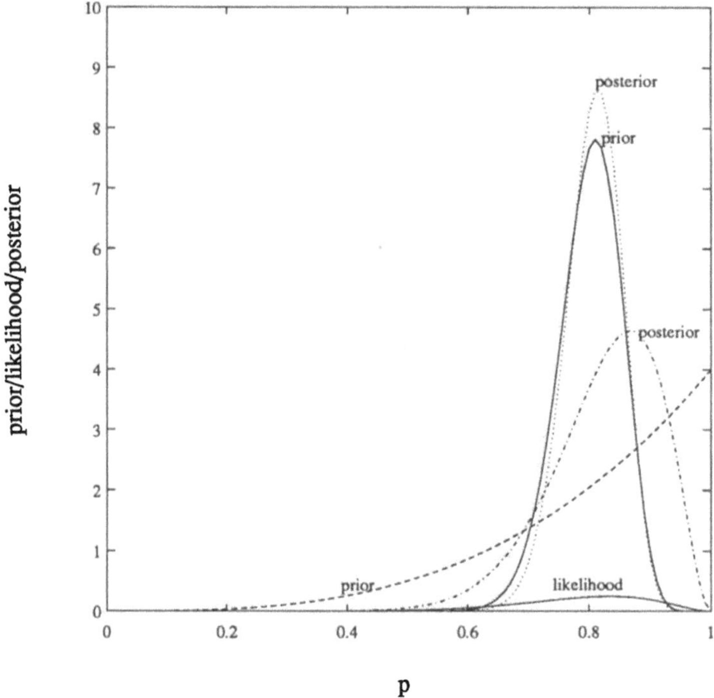

Figure 4. Interaction of priors and likelihood.

It turns out that the methods of the previous sections can be adapted to give estimates of the rate constant λ.

Again the duration of the test, t, is assumed fixed, so the number of failures follows a Poisson law with mean λt (Cox & Lewis (1978)). Thus

$$f(n|\lambda,t) = \frac{(\lambda t)^n \exp\{-\lambda t\}}{n!}$$

In this case the rate, λ, is assumed to have a gamma prior density,

$$g(\lambda;\alpha,\beta) = \frac{\alpha^{\beta}}{\Gamma(\beta)} \lambda^{\beta-1} exp\{-\alpha\lambda\}, \qquad \alpha,\beta > 0,\ \lambda \geq 0$$

The gamma is taken because it is a conjugate prior for the Poisson law. The posterior density is

$$g(\lambda|n) = \frac{(\alpha+t)^{n+\beta}}{\Gamma(n+\beta)} \lambda^{n+\beta-1} exp\{-\lambda(\alpha+t)\}$$

a gamma with parameters α+t and n+β. The rate λ is estimated here by the posterior mean

$$\tilde{\lambda} = E[\lambda|n] = \frac{n+\beta}{\alpha+t}$$

An example will be postponed until the next section which deals more fully with the exponential rate. There it will be seen that t may also be replaced by the total time on test, but the methodology is essentially the same.

3.5. Hazard rate estimation

The arguments of the previous sections may be extended to estimate a hazard rate when the lifetimes are assumed to follow and exponential density. The results, which will not be proved, depend on the interplay between the exponential distribution of times, the Poisson distribution of the number of events in the fixed interval and the gamma waiting time density. The approach is also more general in that it can cope with censored data. In the earlier section only Pascal sampling gave any scope for not requiring the rate of all members of the sample to be known. Thus four test plans are used to cope with some of the possibilities for censorship. In all cases a fixed number, n, of test rigs is assumed.

1. Type II item-censored with replacement

   A predetermined fixed number of failures is determined and the test ends when they have occurred. Failures are replaced as they happen.

2. Type II item-censored without replacement

   Again a fixed number of failures is determined, and testing ends when they have occurred. Failures are not replaced.

3. Type I time-truncated with replacement

   A predetermined test duration is set and the test stops after that time. Failures are replaced.

4. <u>Type I time-truncated test without replacement</u>

Exactly as in 3, but failures are not replaced.

In case 1 and 2 the number of failures, r, is a fixed number and the test duration, T_r, is a random variable. In cases 3 and 4 the number of failures, R, is a random variable and the test duration, t_R, is a fixed number. In both cases denote the ordered failure times by T_i, and these are used to construct the total time on test, TTT. The TTT statistic is in each case (see Martz & Waller (1982)):

1. $TTT = nT_r$

2. $TTT = \sum_{i=1}^{r} T_i + (n-r)T_r \qquad r \leq n$

3. $TTT = nt_R$

4. $TTT = \sum_{i+1}^{R} T_i + (n-R)t_R \qquad R \leq n$

The lifetimes are assumed to follow the density

$$f(t|\lambda) = \lambda\ exp\{-\lambda t\}$$

and the object is to estimate the rate, λ. In all cases the likelihood for the experiment is

$$L = \lambda^m\ exp\{-\lambda TTT\}$$

where m is the observed number of failures. Thus m is r or R depending on the content. The likelihood estimate of λ is easily found as

$$\hat{\lambda} = \frac{m}{TTT}$$

Confidence regions are readily constructed since in both the type II plans the statistic $2\lambda TTT$ has a $\chi^2(2r)$ density, and in the time truncated case with replacement the random number of failures, R, is a Poisson random variable with parameter $n\lambda t_R$.

Moving to the Bayesian approach, the prior $g(\lambda)$ leads to the posterior density

$$g(\lambda|\text{sample}) = \frac{\lambda^m \exp\{-\lambda TTT\} g(\lambda)}{\int_0^\infty \lambda^m \exp\{-\lambda TTT\} g(\lambda) d\lambda}$$

Example 5

100 components are tested for $t_r = 4000$ hrs and R = 2 components are seen to fail. This is Poisson (Type I time truncated with replacement) sampling. Suppose that the prior for λ is the gamma density

$$g(\lambda) = \frac{\alpha^\beta}{\Gamma(\beta)} \lambda^{\beta-1} \, exp\{-\alpha\lambda\}$$

The posterior density is then also the gamma

$$g(\lambda|TTT) = \frac{(\alpha+TTT)^{n+\beta}}{\Gamma(n+\beta)} \lambda^{n+\beta-1} \, exp\{-\lambda(\alpha+TTT)\}$$

and g is estimated by the posterior mean

$$\tilde{\lambda} = E[\lambda|TTT] = \frac{n+\beta}{\alpha+TTT}$$

In this example TTT = 100 x 4000 hr = $4x10^5$ hr, assuming $\beta = 3$, $\alpha = 2.10^5$ hr gives

$$\tilde{\lambda} = E[\lambda|TTT] = \frac{2+3}{2.10^5+4.10^5} = \frac{5}{6.10^5}$$

$$= 0.83x10^{-5} \text{ failures/hr.}$$

The prior is based on assuming a hazard rate of $1.5x10^{-5}$ failures per hour.

References

Cox, D.R. & Lewis, P.A. (1978), *The Statistical Analysis of Series of Events*, Chapman and Hall, London.

LITTLEWOOD, B. & VERRALL, J. (1973), A Bayesian reliability growth model for computer software, *Applied Statistics*, **22**, 332–246.

MANN, N.R., SCHAFER, R.E. & SINGPURWALLA, N.D. (1974), *Methods for the Statistical Analysis of Reliability and Life Data*, John Wiley, New York.

MARTZ, H.J. & WALLER, R.A. (1982), *Bayesian Reliability Analysis*, John Wiley, New York.

FACULTY OF INDUSTRIAL ENGINEERING AND MANAGEMENT SCIENCE
UNIVERSITY OF TECHNOLOGY
P.O.BOX 513
5600 MB EINDHOVEN
THE NETHERLANDS

4. REPAIRABLE SYSTEMS AND GROWTH MODELS

by

MARTIN NEWBY

Eindhoven University of Technology

Abstract

An introduction is given to the Poisson and renewal processes widely used in the study of repairable systems reliability growth, and software reliability. Simple Bayesian treatments of these models are given and it is observed that a full Bayesian treatment requires extensive numerical computations.

1. Introduction

A repairable system is one which is returned to working order after a failure. The questions are then whether the system improves or deteriorates over time, and whether after each repair the system is as good as new. These questions are important in both the development stage and the operational life of a system. When a system is in development the effect of design changes needs to be evaluated, the future cost of the programme requires estimation and a launch date may need to be decided. For an operational system maintenance must be planned and cost monitored and it may be necessary to plan a replacement. In all of these problems changes in performance over time need to be detected and quantified.

Thus for a repairable system the data arrives as a sequence of times of events obtained from each system. Often the data that is available relates only to a single system. This distinguishes the problem from the non-repairable case. The data for the non-repairable case consists of one observation per item on a number of items, whereas for the repairable system there is a number of observations on one item. In non-repairable systems the order of the data is immaterial, but for repairable systems the sequencing of events is critical as is shown in figure 1 (Ascher & Feingold (1984)).

P. Sander and R. Badoux (eds.), Bayesian Methods in Reliability, 101–117.

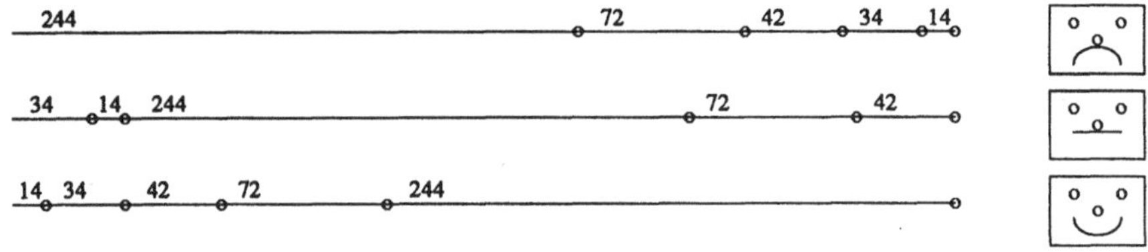

Figure 1. Ascher's happy/neutral/sad example.

The numbers indicate the intervals between failures, and so the same set of intervals clearly show an improving trend, a deteriorating trend and no trend depending on the order in which they are listed. The problem comes down to one of examining sequence of failure times, $T_1 \le T_2 \le T_3 \le \dots \le T_n$ (for convenience define $T_0 = 0$). Moreover, repair times will be ignored in this section. The data may also be recorded as a sequence of inter-failure times $X_1, X_2, \dots, X_n$ where $X_j = T_j - T_{j-1}$ for $j \ge 1$.

Conversely the failure time T_j may be written as $T_j = \sum_{k=1}^{j} X_k$. Along with the times of failures the number of failures in a period [0,t] is of interest. The number of failures will be denoted by N_t, in the following N_t is the counting process associated with the system. It is useful to observe that if $T_n \le t$ then $N_t \ge n$. The time of the n–th event is before t, so there must have been at least n events before t. The process is usually characterised by the probabilities $P(N_t \ge n)$ or $P(T_n \le t)$ or the expected number of failures $E[N_t]$. In view of the remark above, for processes with independent increments

$$P[N_t \ge n] = P[T_n \le t] \tag{1}$$

and

$$P[N_t = n] = P[T_n \le t] - P[T_{n+1} \le t] \tag{2}$$

For simplicity this section will look at two particular models for repairable systems. The names for the models are a little misleading, but are now an established part of the literature.

2. Good as New: the Renewal Process

Failures occur at times T and at each failure the repair returns the system to a new condition. The intervals between failures, X , are independently and identically distributed. A simple example is the replacement of light bulbs, and then the inter-failure times are just the lifetimes of the bulbs. Many simple maintenance problems are described by this model, particularly the situations where the repair consists of simply replacing a failed part.

Under the assumptions the inter-failure times, X_j, are independently and identically distributed with distribution $F(x)$ independent of j and T_j.

Since $T_j = \sum_{i=1}^{j} X_j$ is a sum of independent and identically distributed random variables then the distribution for T_j is $F^{(j)}(t)$, the j-fold convolution of F with itself. By convention $F^{(0)}(t) = 1$. Thus

$$P[T_j \leq t] = P[N_t \geq j] = F^{(j)}(t)$$

on using remark (1).

From (2) it follows that

$$P_n(t) = P[N_t = n] = F^{(n)}(t) - F^{(n-1)}(t)$$

and

$$\begin{aligned} V(t) &= E[N_t] = \sum_{j=1}^{\infty} jP_j(t) \\ &= \sum_{j=1}^{\infty} j[F^{(j)}(t) - F^{(j-1)}(t)] \\ &= \sum_{j=1}^{\infty} F^{(j)}(t) \end{aligned}$$

$V(t)$ is called the renewal function and gives the expected number of failures (or repairs) in time t. While the solution for $V(t)$ is in principle straightforward there are many practical problems in computing $V(t)$. It is also of interest to calculate the *rate of occurence of failures* [ROCOF]

(Ascher & Feingold (1984)), $\frac{dV}{dt}$. For the renewal process there are some simple asymptotic formulae which indicate how the system behaves over a long period. They are

$$\lim_{t\to\infty} \frac{V(t)}{t} = \frac{1}{\mu} \qquad \text{Elementary Renewal Theorem (see Ross (1970))}$$

$$\lim_{t\to\infty} \frac{dV}{dt} = \frac{1}{\mu} \qquad \text{Key Renewal Theorem}$$

$$\lim_{t\to\infty} \frac{N_t}{t} = \frac{1}{\mu} \qquad \text{with probability 1}$$

where μ is the mean inter-failure time, the mean of F. In words the limits above say that over a long period $1/\mu$ replacements per unit time are required. This agrees with intuition.

In discussing repairable systems it is important to be clear about the rate functions that are used. Confusion can arise because in simple models using the exponential density the results are correct even if the rates are confused. The rate of occurrence of failures (often called the failure rate) describes the development of the system over time, measured from zero. The hazard rate (cf. Smith section 4) describes the development only within a failure interval. Figure 1 should make the situation clear, there in each interval the hazard rate repeats itself, while ROCOF is continuous.

The system in Figure 2 has the Erlang density

$$f_{x_i}(x) = xe^{-x}$$

for the intervals, the mean of this density is 2. The discontinuities in the hazard function at each event time are clearly shown. The renewal rate has the simple form

$$v(t) = \tfrac{1}{2} - \tfrac{1}{2}exp\{-2t\}$$

and thus rapidly approaches the asymptotic value of $\frac{1}{2}$.

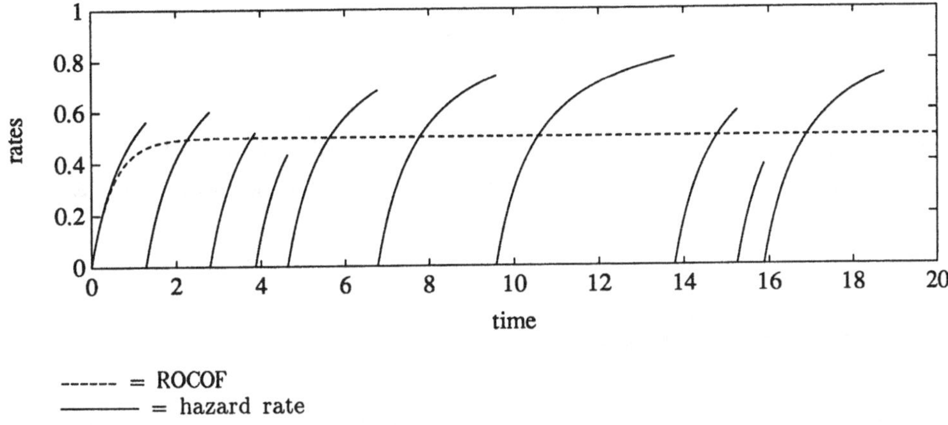

Figure 2. Erlang renewal proces.

3. Estimation

While calculation of the renewal function may be a problem, estimation of the parameters of the inter–failure time distribution F can be dealt with using methods for survival data. Since the inter–failure times, X_j, are independent and identically distributed the sample $(X_1, X_2, ..., X_n)$ may be regarded simply as a sample of size n form the distribution F. The important question remaining is that of deciding whether the process is indeed a renewal process.

4. The Poisson Process

In the simplest renewal process the inter–event times are independent and identically exponentially distributed with density

$$f_x(x) = \lambda e^{-\lambda x}, \; g > 0$$

The number of events has the Poisson distribution with mean λt,

$$P_n(t) = \frac{(\lambda t)^n \; exp\{-\lambda t\}}{n!}$$

$$V(t) = E[N_t] = \lambda t$$

In this case the hazard rate for the interval density is $h(x) = \lambda$ and the rate of occurrence of failures is $v(t) = \lambda$, a source of much confusion.

The Poisson process is relatively easy to handle because the recorded inter-failure times may be treated as a sample from an exponential distribution with mean $1/\lambda$. Testing and estimation can be dealt with exactly as for the failure rate estimation problem.

5. Bad as Old: the Non-Homogeneous Poisson Process

This model again makes strong assumptions about the nature of the process. The key assumptions are that repair times can be neglected, a repair is assumed to be instantaneous and immediately after repair the system behaves like a system of the same age which has experienced no failures. The last assumption is the reason for the name 'bad as old', however it is misleading in that systems which improve fall within the category. Nevertheless the name has become part of the literature.

From the assumptions a failure at time s followed by instantaneous repair means the conditional reliability is $\frac{R(t+s)}{R(s)}$ for a period t into the future. This is just the conditional reliability as if no failure had occured up to time s. It follows quite easily that the failure rate coincides with the hazard rate for the time to first failure and that

$$P[N_t = n] = \frac{\Lambda^n(t)\ exp\{-\Delta t)\}}{n!}$$

and

$$E[N_t] = \Lambda(t)$$

where $\lambda(t)$ is the ROCOF and $\Lambda(t) = \int_0^t \lambda(u)du$

While this model may be challenged, it is simple to deal with and to understand and allows the properties of the system to change over time. Moreover, it is one of the simplest models in which the inter-failure times are not independent and identically distributed.

6. Classical Estimation

If failures are observed at times $(t_0 = 0) < t_1 < t_2 \ldots < t_n$ the likelihood may be written as

$$L = \prod_{i=1}^{n} \lambda(t_i)\, exp\{-[\Lambda(t_i) - \Lambda(t_{i-1})]\}$$

$$= \{\prod_{i=1}^{n} \lambda(t_i)\}\, exp\{-\Lambda(t_n)\} \qquad (3)$$

If a predetermined period of observation [0,T] is chosen then the likelihood is

$$L = \{\prod_{i=1}^{n} \lambda(t_i)\, exp\{-[\Lambda(t_i) - \Lambda(t_{i-1})]\}\}\, exp\{-[\Lambda(T) - \Lambda(t_n)]\} \qquad (4)$$

where the last term is the probability of no events in $(t_n,T]$, as before the exponent forms a collapsing sum and

$$L = \{\prod_{i=1}^{n} \lambda(t_i)\}\, exp\{-\Lambda(T)\}$$

Thus in both cases the likelihood is

$$L = \{\prod_{i=1}^{n} \lambda(t_i)\}\, exp\{-\Lambda(t^*)\}$$

with $t^* = t_n$ or $t^* = T$ as appropriate.

When failure times are not known exactly but are given as grouped data the likelihood is

$$L = \prod_{i=1}^{n} \frac{[\Lambda(t_i)-\Lambda(t_{i-1})]^{f_i}}{f_i!}\, exp\{-[\Lambda(t_i)-\Lambda(t_{i-1})\}$$

where f_i is the number of events in (t_{i-1},t_i).

Thus where a parametric from is assumed classical methods yield estimators for the parameters of interest.

7. Exploratory Analysis

Because $E[N_t] = \Lambda(t)$ plots of the number of failures agains time may indicate the form of the function Λ. This observation forms the basis of Duane analysis and is the reason for the use of such models in reliability growth analysis. Plots of N_t against t fall into three main (see Figure 3) categories (notice that N_t is always a non-decreasing function of t). Where (a) shows reliability growth, (b) a straight line indicating a homogeneous Poisson process, and (c) shows reliability deterioration.

Alternatively plots of N_t/t against time are a measure of the rate of renewals, now N_t/t a horizontal line indicates a homogeneous Poisson process, downward sloping plots show improvement and an upward slope shows deterioration.

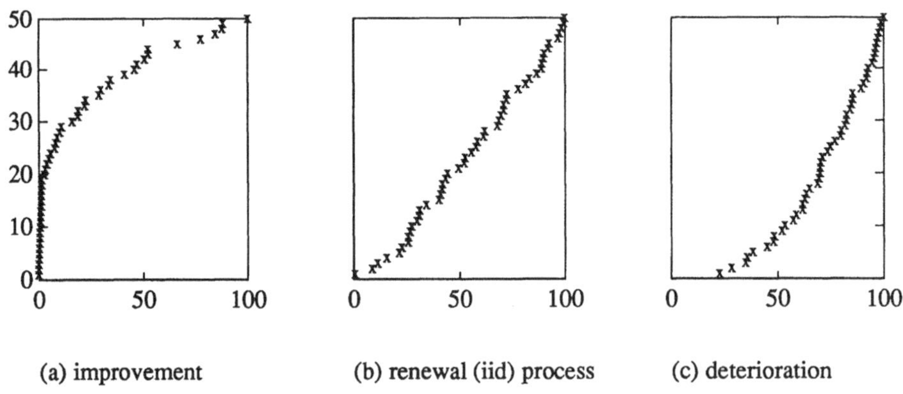

Figure 3. Typical plots of N(t) versus t.

8. The Duane Model

Duane (1964) in a study of a development programme observed that plots of N_t/t were indicative of whether reliability growth is occuring or not. Duane defined N_t/t as the cumulative failure rate (note that N_t/t is neither ROCOF nor a hazard rate) and in many cases found the simple form

$$\frac{N_t}{t} = kt^{-\alpha}, \; k > 0, \; \alpha > 0$$

agreed well with his data. In this case

$$\ln[N_t/t] = \ln(k) - \alpha\ln(t)$$

so that plots of $\ln[N_t/t]$ against $\ln(t)$ will be a straight line.

Crow (1974) formalised the model by replacing N_t by its expectation, $E[N_t]$, in a non-homogeneous Poisson process. Then

$$V(t) = E[N_t] = kt^{1-\alpha}$$

and failure rate

$$v(t) = k(1-\alpha)t^{-\alpha} = (1-\alpha)\frac{V(t)}{t}$$

Crow's Poisson process has $\Lambda(t) = kt^{1-\alpha}$ and $\lambda(t) = k(1-\alpha)t^{-\alpha}$. With this formulation likelihood estimators for k and α follow from (3) or (4) above.

For events recorded at times $t_1 \le t_2 \le ... \le t_n$ or for a time truncated test with $t_1 \le t_2 \le ... \le t_n < T$ the likelihood (from (4) above) is

$$L = [k(1-\alpha)]^n A^{-\alpha} \, exp\{-kt_*^{1-\alpha}\}$$

where $A = \prod_{i=1}^{n} t_i$ and $t_* = t_n$ or $t_* = T$

as appropriate.

The likelihood estimators of α and k are then

$$\hat{\alpha} = 1 - \frac{n}{\sum_{i=1}^{n} \ln(t_*/t_i)}$$

$$\hat{k} = \frac{n}{t_*^{1-\hat{\alpha}}}$$

Balaban (1978) gave the following failure times for an item of equipment.

150	360	541	775	1009	1244	1478	1712
2197	2682	3166	3651	6054	6913	7782	9978
20433	10889	11344	13598	14322	15048	15773	

Table 1. Failure times (in hours)

The Duane plot is given in Figure 4 and the maximum likelihood estimates for k and α are $\hat{k} = 0.04$, $\hat{\alpha} = 0.34$.

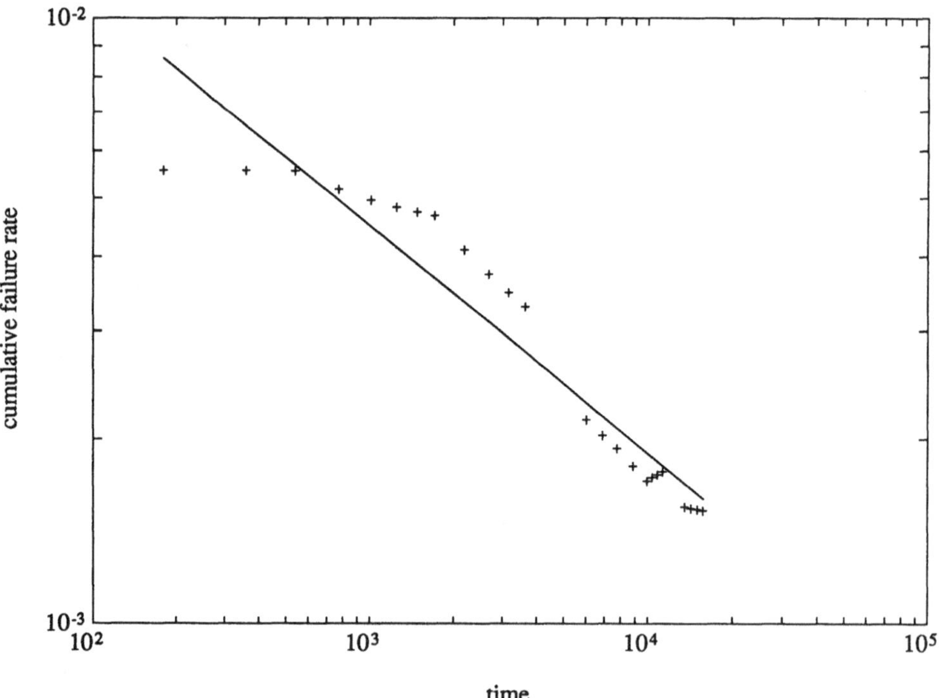

Figure 4. Duane plot for the Balaban data

8. Bayesian Analysis

Only in the simplest cases can Bayesian methods yield results in a closed form. It is almost always necessary to resort to numerical methods to determine the posterior density, its mean or mode. A rather synthetic treatment of the Duane model illustrates the possibilities. The Duane model has two parameters, k and α. Assume that k and α are independently distributed with densities $\psi(k)$ and $\phi(\alpha)$ respectively. The likelihood is

$$L = [k(1-\alpha)]^n A^{-\alpha} \, exp\left\{-kt_*^{1-\alpha}\right\}$$

and thus the posterior density is

$$f(k,\alpha|data) = \frac{k(1-\alpha)^n A^{-\alpha} \, exp\left\{-kt_*^{1-\alpha}\right\}\psi(k)\phi(\alpha)}{\int_0^\infty \int_0^\infty L(k,\alpha|data)\psi(k)\phi(\alpha)dkd\alpha}$$

Two simple cases are $\psi(k)$ a gamma density and $\phi(\alpha)$ a uniform density, and ψ gamma and ϕ deterministic. Suppose then that

$$\psi(k) = \theta(\theta k)^{m-1} e^{-\theta k}/\Gamma(m)$$

and that $\phi(\alpha) = \delta(\alpha_0-\alpha)$. In other words it is known that $\alpha = \alpha_0$. The posterior density then depends only on k and is a gamma

$$f(k|\alpha_0,data) = \frac{\left[\theta+t_*^{1-\alpha_0}\right]^{n-m} k^{n+m-1} \, exp\left\{-k\left[\theta+t_*^{1-\alpha_0}\right]\right\}}{\Gamma(n+m)}$$

The square loss estimator for k is then

$$E[k|\alpha_0,data] = \frac{n+m}{\theta+t_*^{1-\alpha_0}}$$

When

$$\psi(k) = \frac{\theta(\theta k)^{m-1}}{\Gamma(m)} e^{-\theta k}$$

and

$$\phi(\alpha) = \frac{1}{b-a} \qquad a \le \alpha \le b$$

the posterior density is very similar to the likelihood, the integral required cannot be evaluated in closed form, so the best that can be said is

$$f(k,\alpha|\text{data}) = K(1-\alpha)^n A^{-\alpha} k^{n+m-1} \, exp\left\{-k\left[\theta+t^{1-\alpha}\right]\right\} \qquad 0 \le k < \infty,\ a \le \alpha \le b$$

K is the normalising constant.

While the posterior expectation cannot be computed explicitly an alternative estimate is possible. Here the estimators are the modal values of the posterior density. The mode is found by locating the zeros of the derivatives of the posterior density and is therefore a natural extension of the likelihood method. Indeed this method preceded likelihood and was known to Gauss, it is often referred to as the method of inverse probability. The modal values are the solutions of

$$\tilde{k} = \frac{n+m-1}{\theta+t_*^{1-\tilde{\alpha}}}$$

$$\frac{-n}{1-\tilde{\alpha}} - \ln A + \frac{n+m-1}{\theta+t_*^{1-\tilde{\alpha}}} (\ln t_*)\, t_*^{1-\tilde{\alpha}} = 0$$

For ease of calculation a short sequence from a non-homogeneous Poisson process was generated. The process is the Duane/Crow model with k=1, $\alpha=\frac{1}{2}$. The failure times are

0.66	0.78	0.84	1.49	22.66	23.64

Table 2. Failure times for the Duane/Crow model (k=1, α=0.5).

The likelihood function is

$$L = k^6(1-\alpha)^6(345.15)^{-\alpha}\, exp\left\{-k(23.64)^{1-\alpha}\right\}$$

and

$$\hat{k} = 1.41$$

$$\hat{\alpha} = 0.54$$

When α is known to be $\frac{1}{2}$ and k has the gamma density

$$\psi(k) = \frac{3(3k)^2\, e^{-3k}}{\Gamma(3)}$$

the prior mean is 1 and the posterior mean is

$$E[k|\alpha_0,\text{data}] = \frac{6+3}{3+\sqrt{23.64}} = 1.14$$

For the more complex situation take

$$\psi(k) = \frac{3(3k)^2\, e^{-3k}}{\Gamma(3)}$$

and

$$\phi(\alpha) = 1 \qquad\qquad 0 \le \alpha \le 1$$

The posterior density is

$$f(k,\alpha|\text{data}) = K(1-\alpha)^6(345.15)^{-\alpha}\, k^8 exp\left\{-k\left[3+t_*^{1-\alpha}\right]\right\} \qquad 0 \le k < \infty,\ 0 \le \alpha \le 1$$

Here the prior means are $k = 1$, $\bar{\alpha} = \frac{1}{2}$. The modal values of the posterior density are $\tilde{k} = 0.91$, $\tilde{\alpha} = 0.45$. The effects of two priors are shown in Figure 5.

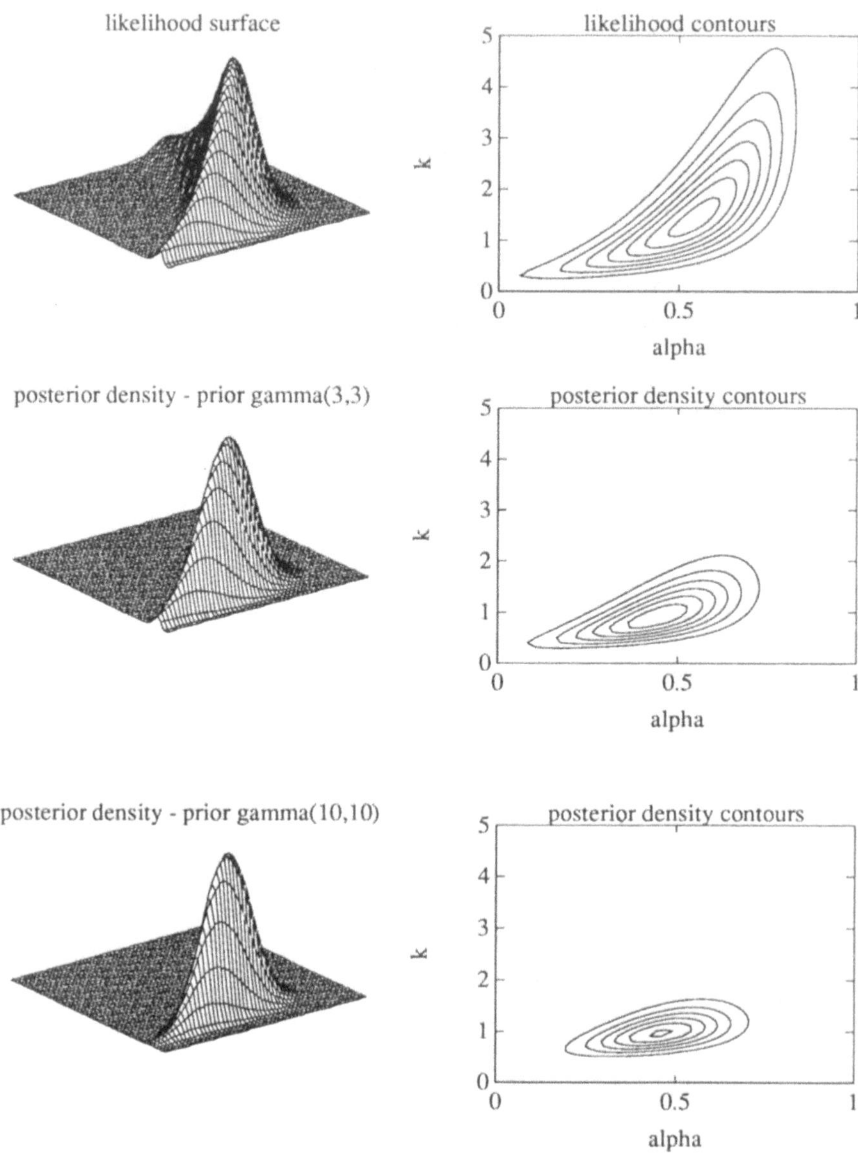

Figure 5. Effect of priors on the Duane model.

The above example serves to show Bayesian methods applied to data from a repairable system. For a more effective application of the techniques a package of numerical methods is required to compute the posterior density and its mean or mode and the appropriate marginal densities (Smith *et al.* (1987), cf. Smith section 18).

A last example illustrates the value of Bayesian and graphical methods in analysing data from repairable systems. In modelling software reliability the initial number of faults in a program is finite. If new faults are not introduced during the life of the program then the mean value function for the process is bounded. This means that standard distribution functions can be used as a basis for the mean value function (Bunday *et al.* (1990)). A Poisson model for software reliability, similar to the Littlewood and Verrall model (Littlewood (1973), cf. Littlewood section 3.4), has a mean value function

$$E[N_t] = \Lambda(t) = \alpha\left\{1 - \frac{1}{(1+t/\beta)^k}\right\} \qquad \alpha,\ \beta,\ k \geq 0$$

where α is the initial number of faults, and the term in braces is a standard Pareto distribution function with parameters β and k (Bunday *et al.* (1990), cf. Newby, chapter 3 section 2.3).

The likelihood function is as in (3) and α can be eliminated using the three likelihood equations to give a profile likelihood for k and β. The profile likelihood is

$$L^*(k,\beta) = K_n \prod_{i=1}^{n} \left\{\frac{k}{(1+t_i/\beta)^{k+1}}\right\} \frac{1}{1 - \dfrac{1}{(1+t_n/\beta)^k}}$$

The profile likelihood and profile log-likelihood for this model applied to the Balaban data are shown in Figure 6. The two functions are very flat, the peak in the likelihood is small, and the contours of the log-likelihood are open to the right. Clearly more information in the form of a prior is required to improve the estimation process.

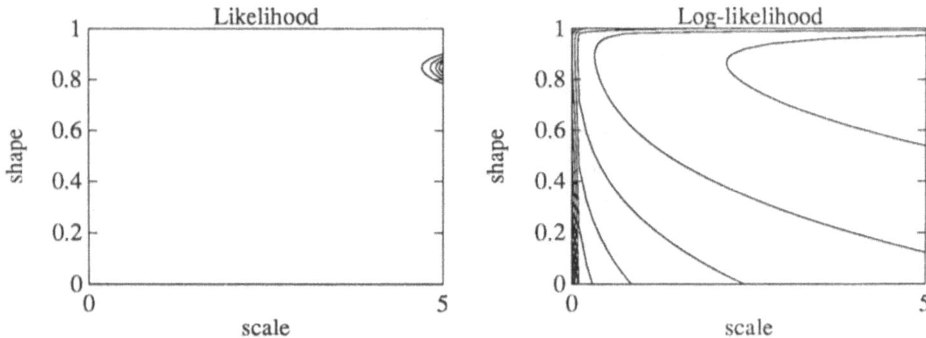

Figure 6. Balaban data analysed with the Pareto model.

References

ASCHER, H. & FEINGOLD, H. (1984), *Repairable Systems Reliability*, Marcel Dekker, New York.

BALABAN, H.S. (1978), Reliability growth models, *Journal of Environmental Science*, 11–18.

BUNDAY, B.D., AL–AYOUBI, I.D. & NEWBY, M.J. (1990), Likelihood and Bayesian Estimation Methods for Poisson Process Models in Software Reliability, *International Journal of Quality and Reliability Management*, **7** (5), 9–18.

CROW, L.H. (1974), Reliability analysis for complex repairable systems. In: F. PROSCHAN and R.J. SERFLING (Eds), *Reliability and Biometry*, 379–410, SIAM Philadelphia.

DUANE, J.T. (1964), Learning curve approach to reliability monitoring, *IEEE Trans. Aero.*, **2**, 563–566.

LITTLEWOOD, B. & VERRALL, J. (1973), A Bayesian reliability growth model for computer software, *Applied Statistics*, **22**, 332–346.

ROSS, S.M. (1970), *Applied Probability Models with Optimization Applications*, Holden–Day Inc., San Francisco.

SMITH, A.F.M., SKENE, A.M., SHAW, J.E.H. & NAYLOR, J.C. (1987), Progress with numerical and graphical methods for practical Bayesian statistics, *The Statistician*, **36**, 75–83.

DEPARTMENT OF INDUSTRIAL ENGINEERING AND MANAGEMENT SCIENCE
UNIVERSITY OF TECHNOLOGY
P.O.BOX 513
5600 MB EINDHOVEN
THE NETHERLANDS

5. THE USE OF EXPERT JUDGEMENT IN RISK ASSESSMENT[1]

by

SIMON FRENCH, FIMA — *University of Leeds*

ROGER M. COOKE — *Delft University of Technology*

MICHAEL P. WIPER — *University of London*

Abstract

Risk assessments must often appeal to expert assessments of failure probabilities. Our concern in this paper is to indicate some of the issues which arise in eliciting and combining expert judgements of the likelihood of particular events or uncertain quantities. We also describe some recent work in the area. In focusing on the mathematical issues, we ignore many others relating, in particular, to the tension between accountability and confidentiality within the management of the risk assessment process.

1. Introduction

The case for conducting risk and reliability analyses during the design stages of many projects and products has been well made. Indeed, such analyses are often mandatory under law in order to show compliance with certain safety criteria. Usually, these analyses may be based upon past failure data and other sources of 'hard' information. However, sometimes sufficient hard data may be unavailable. For instance, in high technology projects using novel materials and components, past failure data sets, particularly in operational environments, are at best sparse and often non-existent. Many systems contain complex embedded software through which there are billions of logically possible paths for control to flow, and thus for which it is impossible to collect sufficient failure data for reliability analyses. In such cases risk assessment can only be based upon the judgements of experts who draw on their knowledge and experience of failures in related but, none the less, substantively different areas.

---

[1] This chapter has been published before in the Bulletin of the Institute of Mathematics and its Applications, 27 (1991), pp. 36-40.

P. Sander and R. Badoux (eds.), Bayesian Methods in Reliability, 119–134.

Our concern in this paper is to indicate some of the mathematical and statistical issues which arise in eliciting and combining expert judgements of the likelihood of particular events or uncertain quantities. We also describe some recent work in the area. At the outset we should be clear that in focusing on the mathematical issues, we ignore many others relating, in particular, to the tension between accountability and confidentiality within the management of the risk assessment process. The interested reader may refer for those to a recent report of the European Safety and Reliability Research and Development Association (1990).

2. Independence Preservation

We begin with a very simple example, originally due to John Pratt and Howard Raiffa (Raiffa (1968)) and further discussed by French in the 1984 IMA conference on 'Analysing Conflict and its Resolution' (French (1987)). This example should serve to indicate that our subject is far from transparent: counter-intuitive results abound and one needs to think very carefully about methodology.

An American salvage contractor is concerned with recovery of a sunken craft from a great depth in the North Sea. Two events that might concern him[2] in thinking about whether to accept the contract are:

A. the £/$ exchange rate rises, and

B. his equipment successfully recovers the craft.

He consults two experts. Both believe that the events A and B are probabilistically independent; and the salvage contractor himself beliefs this too. Thus all concerned agree:

$$P(A \cap B) = P(A)P(B) \qquad (1)$$

However, the first expert believes that the pound is more likely to rise than fall and the recovery more likely to fail than succeed; whereas the other expert believes the pound more likely to fall, but that the recovery is more

[2] Masculine pronouns, etc. are used throughout to indicate persons of either gender

Expert 1	A	A^c	
B	0.14	0.06	0.20
B^c	0.56	0.24	0.80
	0.70	0.30	

Expert 2	A	A^c	
B	0.24	0.56	0.80
B^c	0.06	0.14	0.20
	0.30	0.70	

Table 1: The assessments of the two experts.

(a)	A	A^c	
B	0.19	0.31	0.50
B^c	0.31	0.19	0.50
	0.50	0.50	

(b)	A	A^c	
B	0.25	0.25	0.50
B^c	0.25	0.25	0.50
	0.50	0.50	

(a) probabilities formed by arithmetically averaging the experts' joint probabilities and then marginalising: e.g. 0.19 = (0.14+0.24)/2; then 0.5 = (0.19+0.31).

(b) probabilities formed by arithmetically averaging the experts' marginal probabilities and then forming the joint probabilities to maintain the independence of events A and B: e.g. 0.5 = (0.2+0.8)/2; then 0.25 = 0.5×0.5.

Table 2: Results of averaging the experts assessments arithmetically.

likely to succeed than fail. Specifically they assign joint and marginal probabilities to the events as given in Table 1. Note that the joint probabilities satisfy probabilistic independence (1).

Clearly the experts' beliefs conflict considerably. In assimilating their advice, the contractor must address and resolve this conflict. A simple method might be to average arithmetically the experts' probabilities. However, should the salvage contractor average the joint probabilities or the marginals? Averaging the joint probabilities and then forming the marginal probabilities does not preserve the independence: see Table 2(a). Since all agree events A and B are independent, it would seem that the salvage contractor should proceed as in Table 2(b): he should average the marginals and then form the joint probabilities by enforcing independence.

If the contractor uses geometric averages, the same results are obtained whether he averages the joint probabilities and then forms the marginal probabilities or averages the marginal probabilities and then forms the joint probabilities using independence: see Table 3. (Note, however, that in both cases renormalisation is necessary). Geometric averaging would seem to remove the need for the contractor to make a methodological choice, and might therefore be preferred to arithmetical averaging. Unfortunately, it transpires that any conclusion that we might draw from this example is entirely spurious. Genest & WAGNER (1984) have proved that in general for examples involving fields of more than four events there is no non–dictatorial rule for combining expert judgements that preserves independence. A dictatorial rule simply adopts one expert's judgements without any regard to those of the other experts.

The issue being discussed here is known as the *Principle of Independence Preservation*. This states that if two or more uncertain events or quantities are considered to be probabilistically independent by all the experts then the rule for combining the expert's judgements should preserve this independence. Few principles can seem so uncontroversial and innocuous. Yet Genest and Wagner's result shows that the Principle cannot be achieved in practice. It is perhaps fortunate, therefore, that on deeper reflection the Principle is far from innocuous. If adopted, it would prohibit the design of adaptive combination rules which learn from data about the quality of experts' judgements.

The salvage contractor's statement that A and B are independent means that, if he learnt that A had happened, he would not change his probability for happening. Suppose that, having heard and assimilated the experts' judgements, he learns that A has happened: the exchange rate has risen. The first expert gave A a high probability, whereas the second expert gave it a low one. This might well lead the salvage contractor to give more credence to the advice of the first expert than the second. Thus he might wish to revise his probability

	A	A^c	
B	0.183	0.183	–
B^c	0.183	0.183	–
	–	–	

(a)

	A	A^c	
B	–	–	0.4
B^c	–	–	0.4
	0.458	0.458	

(b)

	A	A^c	
B	0.25	0.25	0.50
B^c	0.25	0.25	0.50
	0.50	0.50	

(c)

(a) The results of geometrically averaging the experts' joint probabilities: $0.183 = \sqrt{(0.14 \times 0.24)} = \sqrt{(0.06 \times 0.56)}$.

(b) The results of geometrically averaging the experts' marginal probabilities: $0.4 = \sqrt{(0.2 \times 0.8)}$; $0.458 = \sqrt{(0.3 \times 0.7)}$.

(c) The same probabilities are obtained whether:

- the four entries in (a) are renormalised to sum to one and then the marginal probabilities formed;
- the row and column marginals in (b) are (separately) renormalised to sum to one and then independence used to form the joint probabilities.

Table 3. Results of averaging the experts assessments geometrically.

for B in order to give more weight to the expert's judgement. Learning that A has occurred leads the contractor to revise his probability for B. For him, once he has assimilated the experts' advice, A and B are no longer independent.

The contractor is not just an observer of two events, A and B; he is also the observer of two forecasting systems on the quality of each of the forecasting systems and, hence, indirectly may affect his probability of the other event by relative weights he gives to the two forecasts.

Since it seems reasonable to revise the relative weights given to different experts or, more generally, different forecasting systems as data is gathered on the quality of their forecasts, we do not adopt the Principle of Independence Preservation in the following.

In passing, we note that Independence Preservation is not the only principle which has been proposed in relation to the combination of experts' judgements and which at first sight seems uncontroversial and innocuous (French (1985).

3. The Quality of Experts' Judgements

Implicit in the above discussion are two points that are central to our development: firstly, it makes sense to talk of the *quality* of an expert's forecast; secondly, one can create the circumstances in which it is possible to gather data which allow the application of adaptive combination rules to bring together the judgements of several experts. In this section we discuss what is meant by the quality of an expert's judgements; in later sections we discuss how adaptive combination rules might be designed.

There has been much research effort expended in investigating the ability of individuals to encode uncertainty consistently (Kahneman *et al.* (1982)). In many senses the results are not encouraging. The predominant view in the literature is that people are not very good at assessing uncertainty. Unaided judgement seems subject to many biases caused by the use of poor heuristics. This conclusion, however, may be suspect because of a further bias: *citation bias*. Typically, papers showing what people cannot do are cited much more frequently than those that show what they can (Beach *et al.* (1987)).

Whatever the case, most of the research has taken place under laboratory conditions. There is evidence that, with training and with the motivation of a real context, many individuals can perform better than they do in laboratory tests (Lichtenstein & Fischoff (1980)). Moreover, it seems that individuals are better at assessing uncertainties in areas in which they have expertise

(Cooke *et al.* (1988 a)).

In assessing the quality of expert advice there are several aspects that must be taken into account.

* **Coherence** Do the judgements cohere: i.e. do the 'probabilities' sum to one and obey all the other relationships that may be expected of them?
* **Temporal consistency** How consistent over time are his judgements? Ideally, his judgements of the same uncertainty made at different times should be identical unless he has reason, such as the acquisition of further evidence, to change them.
* **Calibration** How do the statements of uncertainty relate to observed outcomes? Over a long sequence of predictions do the empirical frequencies of occurrence match the stated probabilities? Thus, if one considers only those occasions on which an expert states a probability of, say, 0.7, do the predicted events happen on 70% of the times.
* **Informativeness** Consider an example. Suppose that each day for a year two forecasters are asked to give their probabilities for rain on the next day. Over the year it rains on 40% of the days: i.e. the base rate of rain is 40%. At the end of the year, it transpires that one forecaster gave his probability as 40% on each and every day. The other gave probabilities of 0% and 100% and perfectly forecast the rainy days. Both are well calibrated; but the second forecaster is the more informative. The information that he gives helps the listener discriminate between wet and dry days in advance. Obviously this example points to two extremes of informativeness. There is a continuum of intermediate possibilities.

An expert's judgement can be checked for coherence at the time that they elicited. Any incoherences found can be brought to his notice for him to reflect on and resolve. Such procedures are invariably a part of the assessment of subjective probabilities (Merkhofer (1987)). Similarly, it is possible, in principle at least, to check for and resolve temporal consistencies should the entire analysis be spread over a significant period of time, although in practice this is seldom done. Hence, in comparing the quality of several experts' judgements, we are mainly concerned with issues of calibration and informativeness.

Several studies have shown that people's judgements of probability are seldom well-calibrated (Lichtenstein *et al.* (1982)). Figure 1 shows the schematic

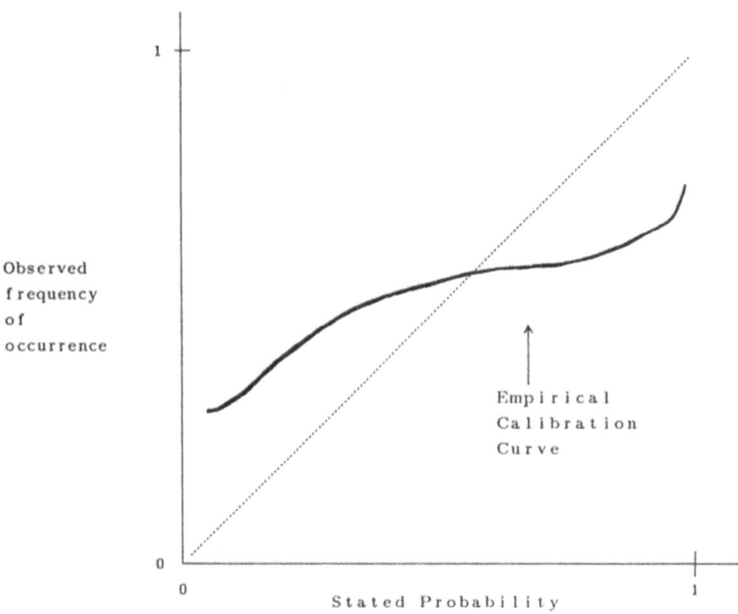

Figure 1. The form of a typical empirical calibration curve

form of an *empirical calibration curve* for judgements concerning the likelihood of events. This is constructed for a given individual by asking him to assess his probabilities for a large number of events. If 60% of the events for which he assessed a probability of 70% actually occur, then his calibration curve would pass through the point (0.7, 0.6). The curve for a well calibrated assessor would simply be the 45° line. It should be emphasised that Figure 1 is schematic: in practice such curves are drawn from fairly limited data sets. More importantly, calibration curves do not indicate how often the individual states particular probabilities: information which is essential for evaluating the significance and quality of their judgements.

If one is to use an expert's judgements, it would seem prudent to calibrate them in some way. There is, however, one further complication. It appears that an individual's calibration depends on the context. An engineer's calibration may be very different on uncertainties within his field of expertise to that on more general uncertainties, say, the chance of rain tomorrow. Thus to recalibrate an expert's judgements for use in a risk analysis, one needs to measure his calibration for the specific context.

Suppose that one recalibrates the judgements of several experts, they do not then become of equal value: some may be more informative than others. In combining their recalibrated judgements, one wishes to give greater weight to the more informative experts.

The discussion in this section has been phrased in terms of judgements concerning uncertain events, for which the expert is simply required to give a single probabilities. In practice, one usually requires experts to give judgements of certain quantities. For these one requires experts to give full probability distributions, or more usually quantiles: say, the median value, the five percentile and the ninety five percentile. The concepts of calibration and informativeness generalise to such cases, but we shall not give the details here (see Cooke (1991)).

4. Calibration Sets and Seed Variables

It is clear from the above that if experts' judgement of uncertainty are to be used in a risk analysis, their calibration and informativeness should be assessed in the context of the analysis. To do this one needs a *calibration set*. A calibration set is a collection of judgements of uncertainty concerning events and quantities the resolution or actual values of which are (or will become) known to the analyst before the experts' judgements that are of real interest are used in the analysis. The uncertain events and quantities in the calibration set are known as *seed variables*. The seed variables should all be drawn from the same context as that of the risk analysis.

For example, suppose that a risk analysis of a manned space–flight requires that the time to notice a change in the reading on a new instrument display is estimated by means of expert judgement. The calibration set for this might involve, *inter alia*, seed variables relating to the times that civil and military pilots take to notice a change in a similar display. Hard data on the latter events may well be obtainable, though not directly known to the experts.

The procedure for using expert judgement, therefore, has the following stages.

1. The events and quantities of real interest are carefully defined.
2. Appropriate calibration sets are constructed for these.
3. The experts are asked for their judgements of uncertainty of the events and quantities of real interest *and* the seed variables in the calibration sets.
4. The quality of the experts is evaluated from their performance on the calibration sets.
5. The experts' judgements of uncertainty of the events and quantities of real interest can combined in the light of what has been learnt about their quality.

For the remainder of the paper we concentrate on the final stage.

5. A Classical Model

Perhaps the most commonly suggested form for combining expert judgement is the *linear opinion pool*:

$$p_c = \sum_e w_e \, p_e \text{ with } \sum_e w_e = 1 \tag{2}$$

where

p_c is the combined probability;
w_e is a nonnegative weight given to expert e;
p_e is the probability judgement of expert e;

and e indexes the experts.

One of the major difficulties that has faced the proponents of opinion pools has been to provide an operational definition of the *weight* of an expert (French (1985)). Clearly, this weight is related to the quality of the expert: but what form should this relation take? Various proposals have been made, although few could be classed as entirely conceptually or operationally sound. In practice, equal weights seem to be used more often than not. Recently, Cooke (1987 and 1991) has developed and justified, both empirically and theoretically, a definition of the weights which is related to the calibration and informativeness of the experts, as assessed from the calibration set. The weight given to expert e has the following form:

$$w_e = \chi_\alpha\{\mathrm{Cal}(e)\} \times \mathrm{Cal}(e) \times \mathrm{Inf}(e) \tag{3}$$

where

Cal(e) is a measure of e's calibration;
Inf(e) is a measure of e's average informativeness;
$\chi_\alpha\{x\}$ is 1 if $x \geq \alpha$, 0 otherwise.

Cooke defines the measure Cal(e) by appealing to the theory of hypothesis testing. If an expert is well calibrated, the actual values of the seed variables give observations drawn from a known distribution. For instance, if a well calibrated expert gives 5%, 50% and 95% quantiles for all the seed variables, then one would expect to see 5% of the actual values of the seed variables falling below their 5% quantile, 45% falling between their 5% and 50% quantiles, etc. In short, one should be observing a sample from a known multinomial distribution. Cal(e) is (a χ^2–approximation to) the probability of observing the actual seed variable values under the hypothesis that he is well calibrated. Thus the better calibrated the expert actually is the larger Cal(e) may be expected to be.

The measure Inf(e) derives from standard information or entropy theory and is taken to be the average relative information of the expert's predictions with respect to uniform distributions. Thus, in a very informal sense, it corresponds to what one would learn on average if one had no information and then learnt the expert's judgement, which one assumed to be well calibrated.

Note the effect of the $\chi_\alpha\{\mathrm{Cal}(e)\}$ term. If an expert is insufficiently calibrated on the seed variables, his weight is set to zero. Hence, an opinion pool formed with these weights includes only the experts who have shown themselves to be sufficiently calibrated and, thus, should be reasonably well calibrated itself. The value of α is chosen to maximise the fit of the predictions given by

$$\sum_e w_e \, p_e$$

to the seed variables. We call the synthesis given by choosing this value of α *the optimal Classical model.*

Cooke (1987) has shown that the weights form ***asymptotic proper scoring rules***: i.e. if an expert wishes to maximise his long run expected weight (influence on the analyst), he will be encouraged to state his true beliefs and lees susceptible pressures that may lead to biases.

6. Bayesian Models

Opinion pools, in a sense, treat the experts' judgements as probabilities and simply seek to form some sort of sensible average. Bayesian methods take the experts' judgements as ***data*** in the light of which the analyst should update his ***a priori*** assessment of the probabilities.

Morris (1974) was the first to introduce a full Bayesian approach to the problem of assimilating experts' judgements. However, only recently have Bayesian methods been developed for learning about the relative quality of experts from a calibration set. Mendel & Sheridan (1987) first developed a model that attempted to do so. Unfortunately its assumptions are somewhat dubious (Cooke (1991)) and, moreover, it is not computationally feasible. French & Wiper (1990) have developed an alternative model, arguably based on equally dubious assumptions, but at least one that is computationally feasible.

The details of this model are complex, but it is appropriate to indicate the difficulty that Bayesian methods face. The seed variables in a calibration set are seldom measured on a common scale. Some seed variables, for instance, may relate to times between successive failures of a component; others to the frequency with which certain circumstances are encountered. Nonparametric models can avoid the need to bring all the seed variables to a common scale; but Bayesian non-parametric models are very difficult computationally. Thus one is driven to parametric models, which require that all seed variables are transformed to a common scale. French and Wiper achieve this by transforming all the experts' judgements to quantiles of the analyst's prior distributions. This makes their model sensitive to the analyst's prior distributions, although a careful use of non-informative priors seems to minimise this. None the less, a lot of further work needs to be done.

In principle, Bayesian models should have a strong practical advantage over the Classical model. They seek to ***recalibrate*** the experts' judgements and to draw upon all the available information. The Classical model discards the judgements of poorly calibrated experts, and thus only draws upon some of the available information.

7. Some Experimental Results

The Classical model has been tested in a variety of contexts and has performed successfully in all but a few cases. It appears to have the robustness that one requires from an operational method. The Bayesian model, being a year or two younger, is less well tested and can at best be described as promising. Here we describe their application to a data set concerned with predicting the reliability of flanges (Cooke *et al.* (1988 b)).

Ten experts were asked to state 5%, 50% and 95% quantiles for a set of eight seed variables and six variables of real interest, concerned with irregularities in flanges and the hazard rates of flanges. The experts comprised operators, mechanics, maintenance engineers and their supervisors with experience of the problem of flange failure.

In evaluating the models, we concentrate on the eight seed variables since it is only for all those that we know the actual values. Table 4 gives the calibration and information scores of each of the ten experts, as well as the product of these, $\mathrm{Cal}(e)\times\mathrm{Inf}(e)$. This is the weight given in (3) stripped of the $\chi_\alpha\{\mathrm{Cal}(e)\}$ term. We have found this quantity a very good indicator of the quality of an expert; at least in so far as a single number can represent a complex multi-faceted quality. We also give results for four syntheses of the experts' judgements: i.e. we evaluate the four syntheses in terms of their calibration, information and $\mathrm{Cal}(e)\times\mathrm{Inf}(e)$ over the seed variables. In doing this, we realise that we are overfitting our limited data set. However, there is little else that we can do. None the less, absolute values should be ignored and only relative ordering of the quantities considered.

From Table 4 it is clear that eight of the ten experts are very poorly calibrated. Indeed, the term $\chi_\alpha\{\mathrm{Cal}(e)\}$ in the weights used in the Classical model is zero for all but experts 2 and 8. Including all ten experts with equal weights gives a much worse performance than the optimal Classical model. The Bayesian models apparently out perform the Classical, even when restricted to using experts 2 and 8 only: i.e. the same data as the Classical. However,

	Calibration	Information	Cal(e)×Inf(e)
Expert 1	.004	1.227	.004
Expert 2	.541	.920	.497
Expert 3	.004	1.672	.006
Expert 4	.027	.922	.024
Expert 5	.004	1.227	.004
Expert 6	.000	1.510	.000
Expert 7	.003	1.586	.005
Expert 8	.145	1.205	.174
Expert 9	.000	1.842	.000
Expert 10	.027	1.824	.048
Equal weights (a)	.185	1.037	.192
Classical (b)	.664	.943	.626
Bayesian (c)	.664	1.432	.950
Bayesian (d)	.640	2.186	1.400

(a) The linear opinion pool for all ten experts using equal weights.
(b) The Classical model, which uses experts 2 and 8 only.
(c) A Bayesian model applied to experts 2 and 8 only.
(d) A Bayesian model applied to all ten experts.

Table 4. Results evaluated in terms of Cal(e)×Inf(e).

this should be taken with a degree of scepticism. There are many distributional and parameter choices to be made in implementing the Bayesian model. These have been made in full knowledge of the properties of this data set. It is not surprising that the models perform well! What we claim is that the Bayesian models are showing promise worthy of greater study. It should also be noted that the Bayesian model is producing distributions which are arguably over-confident in their informativeness. Again further research is necessary. For further details of the Bayesian model see French & Wiper (1990) and Wiper (1990).

References

BEACH, L.R., CHRISTENSEN-SZALANSKI, J. & BARNES, V. (1987). Assessing human judgement: has it been done, can it be done, should it be done? In: G. WRIGHT & P. AYTON (Eds), *Judgemental Forecasting*, John Wiley, Chichester, 49 – 62.

COOKE, R.M. (1987). A theory of weights for combining expert opinion. Department of Mathematics, Delft University of Technology.

COOKE, R.M. (1991). *Experts in Uncertainty: Expert Opinion and Subjective Probability in Science.* Oxford University Press.

COOKE, R.M., MENDEL, M.B. & THEYS, W. (1988 a). Calibration and information in expert resolution: a classical approach, *Automatica*, **24**, 87 - 94.

COOKE, R.M., STOBBELAAR, M. & Van Steen, J. (1988 b). Expert Opinion in Safety Studies:Case Report 4 - DSM Case. Department of Mathematics, Delft University of Technology.

European Safety and Reliability Research and Development Association (1990). *Expert Judgement in Risk and Reliability Analysis: Experience and Perspective.* ESSRDA Report No. 2.

FRENCH, S. (1985). Group consensus probability distributions: a critical survey. In: J.M. BERNARDO, M.H. DEGROOT, D.V. LINDLEY & A.F.M. SMITH (Eds), *Bayesian Statistics* 2, North Holland, Amsterdam, 183 - 201.

FRENCH, S. (1987). Conflict of belief: when advisers disagree. In: P.G. BENNETT (Ed), *Analysing Conflict and its Resolution*, Oxford University Press.

FRENCH, S. & WIPER, M.P. (1990). Bayesian Updating by remodelling an Expert's Quantiles. Report 90.7, School of Computer Studies, University of Leeds.

GENEST, C. & WAGNER, C. (1984). Further evidence against independence preservation in expert judgement synthesis.

KAHNEMAN, D., SLOVIC, P. & TVERSKY, A. (Eds) (1982). *Judgement under Uncertainty: Heuristics and Biases*, Cambridge University Press.

LICHTENSTEIN, S. & FISCHOFF, B. (1980). Training for calibration, *Organisational Behaviour and Human Performance*, **28**, 149 - 171.

LICHTENSTEIN, S., FISCHOFF, B. & PHILLIPS, L.D. (1982). Calibration of probabilities: the state of the art until 1980. In: KAHNEMAN *et al*, 306 - 334.

MENDEL, M.B. & SHERIDAN, T. (1987). Optimal estimation using human experts. Man-machine Systems Lab., Department of Mechanical Engineering, MIT.

MERKHOFER, M.W. (1987). Quantifying judgemental uncertainty: methodology experiences and insights, *IEEE Transactions on Systems, Man and Cybernetics*, **17**, 741 - 752.

MORRIS, P.A. (1974). Decision analysis: expert use, *Management Science*, **20**, 1233 - 1241.

RAIFFA, H. (1968). *Decision Analysis*. Addison Wesley, Reading, Mass.

WIPER, M.P. (1990). *Calibration and Use of Expert Probability Judgements*. PhD Thesis, School of Computer Studies, University of Leeds.

SCHOOL OF COMPUTER STUDIES
UNIVERSITY OF LEEDS
MATHEMATICAL SCIENCES
LEEDS, LS2 9JT
UK

DEPARTMENT OF MATHEMATICS
DELFT UNIVERSITY OF TECHNOLOGY
JULIANALAAN 132
2628 BL DELFT
THE NETHERLANDS

DEPARTMENT OF MATHEMATICAL STUDIES
GOLDSMITHS' COLLEGE
UNIVERSITY OF LONDON
NEW CROSS
LONDON, SE14 6NM
UK

6. FORECASTING SOFTWARE RELIABILITY [1]

by

BEV LITTLEWOOD

City University, London

Abstract

Computer software fails because of the presence of intellectual faults, ranging from simple coding faults to fundamental design faults. In principle, such faults can be permanently removed when they are detected by failure of the software. Then the software will exhibit reliability growth. The problem considered here is the one of forecasting this growth: it includes the estimation of the current reliability of the program from the previous failure data. We begin with a brief description of the software failure process: a non-stationary stochastic process. Several of the best-known software reliability growth models are described, and examples given of their performance on real software failure data. They shown marked disagreement and thus reveal a need for methods of comparing and evaluating software reliability forecasts. Several simple techniques for conducting this evaluation are described and illustrated using several different models on real data sets. Finally, it is shown how in certain circumstances it is possible to improve the predictive accuracy of software reliability models by a re-calibration technique.

1. Introduction

Software reliability models first appeared in the literature almost fifteen years ago (Jelinski & Moranda (1972), Shooman (1973), Littlewood & Verrall (1973), Musa (1975)), and according to a recent survey some forty now exist (Dale (1982)). There was an initial feeling that a process of refinement would eventually produce definitive models which could be unreservedly recommended

---

[1] This chapter has been published before, see Littlewood, B (1989), Forecasting Software Reliability, Lecture Notes in Computer Science, No. 341, Springer-Verlag, Berlin.

P. Sander and R. Badoux (eds.), Bayesian Methods in Reliability, 135–201.

to potential users. Unfortunately this has not happened. Recent studies suggest that the accuracy of the models is very variable (Keiller *et al* (1983 b)), and that no single model can be trusted to perform well in all contexts. More importantly, it does not seem possible to analyse the *particular* context in which reliability measurement is to take place so as to decide a priori which model is likely to be trustworthy.

Faced with these problems, our own research has recently turned to the provision of tools to assist the user of software reliability models. The basic device we use is an analysis of the predictive quality of a model. If a user knows that past predictions emanating from a model have been in close accord with actual behaviour for a particular data set then he/she can have confidence in future predictions for the same data.

We shall describe several ways of analysing predictive quality, generally of a fairly informal nature. These techniques will be illustrated using several models to analyse several data sets. Our intention, however, is not to act as advocates for particular models, although some models do seem to perform noticeably more badly than others. Rather, we hope to provide the beginnings of a framework which will allow a user to have confidence in reliability predictions calculated on an everyday basis.

An important by-product of our ability to analyse predictive quality will be methods of improving the accuracy of predictions. We shall show some remarkably effective techniques for obtaining better predictions than those coming from 'raw' models, and suggest ways in which other 'meta' predictors might be constructed.

2. The Software Reliability Growth Problem

The theme of this paper is prediction: how to predict and how to know that predictions are trustworthy.

We shall restrict ourselves, for convenience, to the continuous time reliability growth problem. Tables 1, 2, 3 show typical data of this kind. In each case the times between successive failures are recorded. Growth in reliability occurs as a result of attempts to fix faults, which are revealed by their manifestation as failures.

3	30	113	81	115	9	2	91	112	15
138	50	77	24	108	88	670	120	26	114
325	55	242	68	422	180	10	1146	600	15
36	4	0	8	227	65	176	58	457	300
97	263	452	255	197	193	6	79	816	1351
148	21	233	134	357	193	236	31	369	748
0	232	330	365	1222	543	10	16	529	379
44	129	810	290	300	529	281	160	828	1011
445	296	1755	1064	1783	860	983	707	33	868
724	2323	2930	1461	843	12	261	1800	865	1435
30	143	108	0	3110	1247	943	700	875	245
729	1897	447	386	446	122	990	948	1082	22
75	482	5509	100	10	1071	371	790	6150	3321
1045	648	5485	1160	1864	4116				

Table 1. Execution times in seconds between successive failures (Musa (1979)). Read left to right in rows.

479	266	277	554	1034	949	693	597	117	170
117	1274	469	1174	693	1908	135	277	596	757
437	2230	437	340	405	575	277	363	522	613
277	1300	821	213	1620	1601	298	874	618	2640
5	149	1034	2441	460	565	1119	437	927	4462
714	181	1485	757	3154	2115	884	2037	1481	559
490	593	1769	85	2836	213	1866	490	1487	4322
1418	1023	5490	1520	3281	2716	2175	3505	725	1963
3979	1090	245	1194	994	3902				

Table 2. Execution time in hundredths of seconds between successive failures.

39	10	4	36	4	5	4	91	49	1
25	1	4	30	42	9	49	44	32	3
78	1	30	205	5	129	103	224	186	53
14	9	2	10	1	34	170	129	4	4
35	5	5	22	36	35	121	23	33	48
32	21	4	23	9	13	165	14	22	41
12	138	95	49	62	2	35	89	90	69
22	15	19	42	14	11	41	210	16	30
37	66	9	16	14	24	12	159	89	118
29	21	18	2	114	37	46	17	1	150
382	160	66	206	9	26	62	239	13	4
85	85	240	178	34	102	9	146	59	48
25	25	111	5	31	51	6	193	27	25
96	26	30	30	17	320	78	39	13	13
19	128	34	84	40	177	349	274	82	58
31	114	39	88	84	232	108	38	86	7
22	80	239	3	39	63	152	63	80	245
196	46	152	102	9	228	220	208	78	3
83	6	212	91	3	10	172	21	173	371
40	48	126	90	149	30	317	500	673	432
66	168	66	66	120	49	332			

Table 3. Operating times between successive failures. This data relates to a system experiencing failures due to software faults and hardware design faults.

We shall take a program, **p**, to be a mapping **p**: **I** ⟶ **O**. Here **I** is the input space, i.e. the totality of all possible inputs, and **O** is the output space. The failure process is then illustrated in Figure 1.

In Figure 1(a) we see that there are certain inputs which the program **p** cannot exercise correctly. These comprise the subset $\mathbf{I}_F$ of all inputs **I**. In practice a failure will be detected by a comparison between the output obtained by processing a particular input, and the output which ought to have been produced according to the specification of the program. Detection of failures is, of course, a non-trivial task, but we shall not concern ourselves with this problem here.

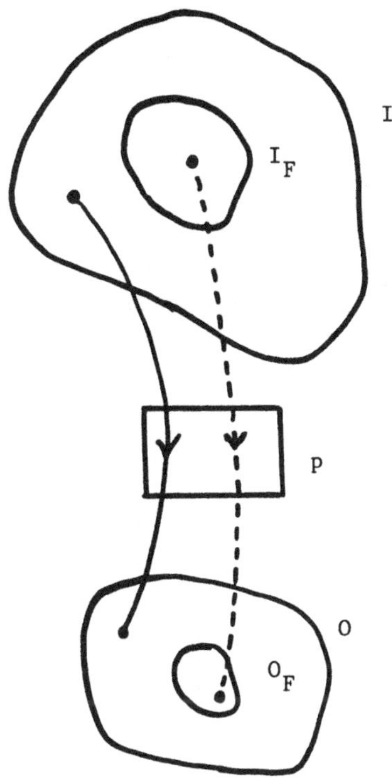

Figure 1(a)

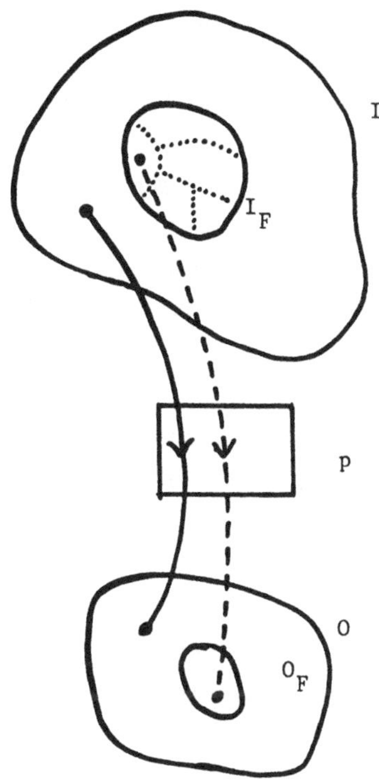

Figure 1(b)

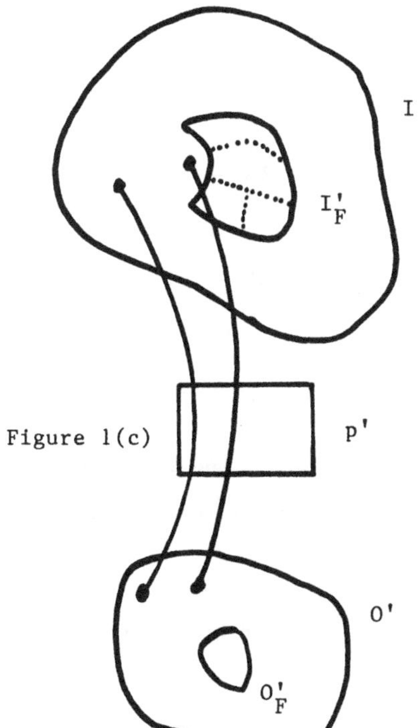

Figure 1(c)

Figure 1. A conceptual model of the software failure process as a mapping **p**: **I** → **O**. Here broken lines indicate "failed" mappings.

Here **O** can be taken to be the set of all outputs which can be produced by the processing of all the possible inputs represented by **I**. The subset of failure-prone inputs $\mathbf{I}_F$ produces the subset $\mathbf{O}_F$ of failed outputs.

We can take the conceptual model a stage further by considering the underlying *faults* which reside in the program **p**, Figure 1(b). If we make the reasonable assumption that each failure can be said to have been caused by one (and only one) fault, we have a partitioning of $\mathbf{I}_F$ into subsets corresponding to the different faults.

When we successfully remove a fault, and so change the program **p** into a new program **p'** (see Figure 1 (c)), this has the effect of removing certain points of **I** from $\mathbf{I}_F$. Thus the members of the removed fault set now map into "acceptable" regions of **O**.

Operational use of a program may be thought of as the selection of a trajectory of points in the space **I**. Typically many inputs will be successful, i.e. outside $\mathbf{I}_F$, before an input is selected which lies in $\mathbf{I}_F$ and so causes a failure. When the failure occurs, an attempt will be made to fix the underlying fault and if this attempt is successful we have the situation shown in the transition from Figure 1(b) to Figure 1(c).

Execution of the program then restarts (most probably in a region outside $\mathbf{I}_F$, since $\mathbf{I}_F$ is typically very small), and the trajectory of successive inputs continues until the next failure when the fixing operation is repeated.

The result is a sequence of programs $\mathbf{p}_1, \mathbf{p}_2, \mathbf{p}_3, \ldots, \mathbf{p}_n, \ldots$; a sequence of successively smaller sets $\mathbf{I}_F^{\,1}, \mathbf{I}_F^{\,2}, \mathbf{I}_F^{\,3}, \ldots, \mathbf{I}_F^{\,n}, \ldots$; a sequence of output sets $\mathbf{O}^1, \mathbf{O}^2, \mathbf{O}^3, \ldots, \mathbf{O}^n, \ldots$; and a sequence of subsets of failed outputs $\mathbf{O}_F^{\,1}, \mathbf{O}_F^{\,2}, \mathbf{O}_F^{\,3}, \ldots, \mathbf{O}_F^{\,n}, \ldots$. Clearly the reliability growth is determined by the sequence $\{ \mathbf{I}_F^{\,i} \}$.

In this paper we shall confine ourselves to the continuous time case. There is a sense, of course, in which the whole problem is really a discrete time one (computer systems are discrete devices; our conceptual model relies on the idea of discrete input cases). However, the times between successive failures, which will be the random variables of interest, will typically be very much larger than the machine cycle times and the times required to process individual inputs. Therefore a continuous time approach will be a good approximation to what is really happening.

With this proviso, it seems reasonable to assume that the sets I_F are encountered purely randomly in the execution trajectory. That is, the time to next failure (and so the inter-failure times) has, conditionally, an exponential distribution. If we let $T_1, T_2, T_3, \ldots, T_n, \ldots$ be the successive inter-failure times, we have a complete description of the stochastic process if we know the rates $\lambda_1, \lambda_2, \lambda_3, \ldots, \lambda_n, \ldots$.

Clearly these rates, and in particular the successive differences representing the improvements caused by the attempts to remove faults, will depend on the "sizes" of the subsets representing the faults in I_F. There will be a tendency for the larger faults to be detected, and so removed, earlier: this implies a law of diminishing returns for debugging. However, the failure subset I_F will be encountered randomly and so will its subsets corresponding to the faults. There is no guarantee that faults will be encountered in order of their size. In fact the sequence of successive fault sizes is a stochastic process, and so therefore is the corresponding sequence of rates corresponding to $T_1, T_2, T_3, \ldots$. Call the latter $\Lambda_1, \Lambda_2, \Lambda_3, \ldots$.

As this model stands, we would expect that $\Lambda_1 > \Lambda_2 > \Lambda_3 > \ldots$. However, we have so far assumed that a fix is certain to be effective, and therefore the only uncertainty concerns its magnitude. In fact this may be unrealistic. There is evidence that fix attempts are fallible and that sometimes a program is made less reliable as a result of an attempt to remove a fault. It might be more realistic, therefore, merely to insist that the $\{\Lambda_i\}$ sequence is *stochastically* decreasing.

To summarise this conceptual model so far: there are two sources of uncertainty in the failure behaviour of software which is being debugged. In the first place, there is uncertainty arising from the *operational environment*: specifically in the sequence of input cases selected for processing by the program. Even if we knew I_F, we would not know when an input would next be selected from here. This uncertainty results in the assumption of conditional exponential distributions for the inter-failure times. Secondly, there is uncertainty arising from the *debugging operation* itself. Even if we knew the partitioning of I_F, we would not know which fault would be encountered next and so we would be uncertain about the magnitude of the change in the failure rate. This results in the sequence of rates (parameters of the exponential distributions of the T's) being a sequence of random variables: that is we have a doubly stochastic scheme.

Not all the reliability growth models for software in the literature follow this doubly stochastic approach. In this paper we shall describe the most popular models and indicate their relationship to our conceptual model. It should be pointed out, though, that this model has some serious limitations. We have tended to gloss over the exact nature of the *failure set* $\mathbf{I}_F$, and the *trajectories* of successive inputs in operational use. Figure 1 shows $\mathbf{I}_F$ as a connected set in some unspecified topology. It is unlikely that this is ever the case: different faults will often be "far away" from one another, separated by tracts of failure free inputs. It is even possible that single faults do not constitute connected sets. We know very little about the properties of these fault sets, but some preliminary experimental work suggests that software faults can have very strange shapes (Amman & Knight (1987)). Even less is known about the trajectories of successive inputs corresponding to operational use. It seems likely that in some applications, such as process control (chemical plant, fly-by-wire avionics), the successive inputs will be very close and the trajectory will look like a (fairly) random walk in **I**. If the trajectory is fairly smooth in this way, and if the individual fault sets are connected, then we could expect to see clusters of failures. A trajectory will wander around **I** until it hits a fault and then it is likely that a sequence of successive inputs will all lie in the same fault set. This could have serious implications for the safety of a fly-by-wire aircraft: the internal stability of the physical system could probably tolerate individual failed cycles of the control system but not a cascade of such failures. In what follows it should be assumed that we are not dealing with such situations. Rather there is an assumption that, following a failure and fix attempt, the software is restarted at a point in **I** which is "distant" from the point at which it failed. This is realistic in many applications.

Before we come to brief descriptions of some models of this reliability growth process, it is important to consider just what is our objective. The basic problem is shown in Figure 2. The raw data available to the user will be a sequence of execution times t_1, t_2, ..., t_{i-1}, between successive failures. These observed times can be regarded as realisations of random variables T_1, T_2, ..., T_{i-1}. The objective is to use the data, observations on the past, to

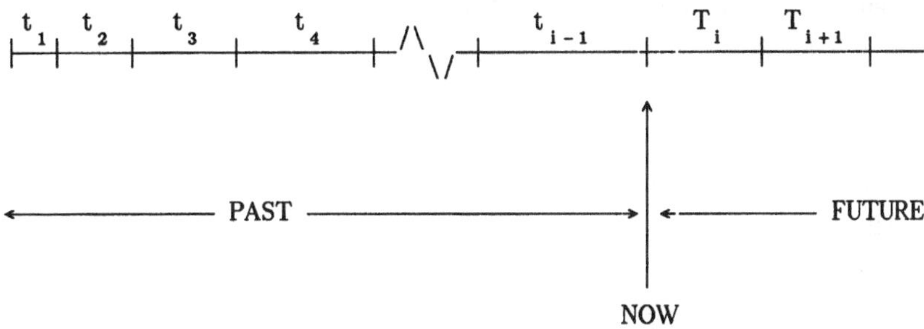

Figure 2. The problem is to make predictions now about the future. using only data collected in the past

predict the future unobserved T_i, T_{i+1}, ... It is important to notice that even the simplest problem concerning measurement of *current reliability* is a prediction: it involves the future via the unobserved random variable T_i.

It is this characterisation of the problem as a *prediction problem* which will underlie all our work reported in this paper. We contend that the only important issue for a user is whether future behaviour can be accurately predicted. Other metrics, such as estimates of the number of faults left in a program, are of interest only inasmuch as they contribute to this overall aim of predicting with accuracy. Indeed, a recent study (Adams (1984)) of several large IBM systems suggests that this particular metric can be very misleading: systems with very many faults can have acceptably high reliability if each fault occurs very infrequently.

Informally, the prediction problem is solved if we can accurately estimate the joint distribution of any finite subset of T_i, T_{i+1}, ... This statement begs the question of what we mean by 'accurately', and it is this issue which forms a major part of our work.

In practice, of course, a user will be satisfied with much less than a complete description of all future uncertainty. In many cases, for example,it will be sufficient to know the current reliability of the software under examination. This could be presented in many different forms: the reliability function, $P(T_i > t)$; the current rate of occurrence of failures (ROCOF),(cf. Newby chapter 4 section 2 and Ascher & Feingold (1984)); the mean (or median) time to next failure (MTTF). Alternatively, a user may wish to predict when a *target reliability*, perhaps used as a criterion for a test termination, will be achieved.

If we accept that prediction is our goal, it can be seen that the usual discussion of competing software reliability growth *models* is misleading. We should, instead, be comparing the relative merits of *prediction systems*. A prediction system which will allow us to predict the (T_i, T_{i+1}, ...) from the past (t_1, t_2, ..., t_{i-1}) comprises:

(i) the *probabilistic model* which specifies the distribution of any subset of the T_j's conditional on a (unknown) parameter α;

(ii) a statistical inference procedure for α involving use of available data (realisations of T_j's);

(iii) a *prediction procedure* combining (i) and (ii) to allow us to make probability statements about future T_j's.

Of course, the *model* is an important part of this triad and it seems unlikely that good predictions can be obtained if the model is not 'close to reality'. However, a good model is not sufficient: stages (ii) and (iii) are vital components of the prediction system. In fact disaster can strike at any of the three stages.

In principle, it ought to be possible to analyse each of the three stages separately so as to gain trust in (or to mistrust) the predictions. Unfortunately, it is our experience that this is not possible. There are several reasons.

In the first place, the *models* are usually too complicated for a traditional 'goodness-of-fit' approach to be attempted. Even the simplest exponential order statistic model (Jelinski & Moranda (1972), Shooman (1973), Musa (1975)) does not allow this kind of analysis. This should not surprise us: the goodness-of-fit problem for independent *identically* distributed random variables is hard in the presence of unknown parameters. The reliability growth context is much worse because of non-stationarity.

Secondly, properties of the estimators of unknown parameters for a non-Bayesian analysis of these models are usually not available. For example, several models assume that the software contains only a finite number of faults. There is thus an upper bound on the number of observable T_j's. This implies that we cannot even trust the usual asymptotic theory for maximum likelihood (ML) estimators. Their small sample properties are invariably impossibly hard to obtain.

Of course, there is a proper approach to stages (ii) and (iii) in the Bayesian framework. It involves posterior distributions of the parameters at stage (ii) and Bayesian predictive distributions for (iii) (cf. Smith section 15 and Aitchison & Dunsmore (1975)). Unfortunately, this does present some analytical difficulties for the popular software reliability growth models. However, with recent advances in Bayesian numerical techniques, coupled with powerful personal computers, this picture may change in the near future.

Finally, it could be argued that there are models which are 'obviously' better than others because of the greater plausibility of their underlying assumptions. We find this a rather dubious proposition. Certainly, the assumptions of some models seem overly naive and it might be reasonable to discount them. However, this still leaves others which cannot be *rejected a priori*. It is our belief that understanding of the processes of software engineering is so imperfect that we cannot even choose an appropriate model when we have an intimate knowledge of the software under study. At some future time it may be possible to match a reliability model to a program via the characteristics of that program: for example using complexity metrics. This is not currently the case.

Where does this leave a user, who merely wants to obtain trustworthy reliability metrics for his current software project? Our view is that there is no alternative to a direct examination and comparison of the quality of the predictions emanating from different complete prediction systems. The intention in this paper is to present the beginnings of a set of tools to assist this examination and comparison. Although we shall show examples of these tools applied to the predictions from several prediction systems using real software reliability data, our intention is *not* to recommend particular ways of predicting. Rather, it is our experience that no prediction system can be trusted to be always superior to others. Our advice to users, then, is to be eclectic: try many prediction systems and use the reliability metrics which are the best *for the data under consideration*.

We would not claim that the tools described here are complete. Indeed, we shall suggest potentially fruitful areas for future research. We do believe strongly, however, that now is an appropriate time to devote less effort to the proliferation of models and more to studies of this kind.

3. Some Software Reliability Growth Models

In this section we shall describe briefly some software reliability prediction systems. The models described here are only a small subset of those which appear in the literature: see Dale (1982) for a near exhaustive survey.

3.1. Jelinski and Moranda (JM)

This model (Jelinski & Moranda (1972)) is justifiably credited with being the first reliability growth model specifically created for software. The model due to Shooman (1973) appears to be identical to JM and apparently arises from independent work at approximately the same time. The Musa model (Musa(1975)) introduces extra refinements, and has been fairly widely used, but its foundations are essentially identical to JM.

The JM model assumes that T_1, T_2, ... are independent random variables with exponential probability density functions

$$p(t_i|\lambda_i) = \lambda_i e^{-\lambda_i t_i} \qquad t_i > 0 \tag{1}$$

where

$$\lambda_i = (N-i+1)\phi \tag{2}$$

The rationale for the model is as follows. When our observation of the reliability growth (debugging) begins, the program contains N faults. Removal of a fault occurs whenever a failure occurs, and at each such event the rate of occurrence of failures is reduced by an amount ϕ . Thus ϕ can be taken to represent the size of a fault.

Another way to consider the model is via competing risks. All faults in a program can be considered to be waiting for discovery. Their discoveries will occur at times X_1, X_2, ..., X_n, ... (for some arbitrary labelling of the faults) measured from the beginning of debugging. It is assumed that all faults are similar, so the X_i can be treated as identically distributed. It is further assumed (quite reasonably) that the X_i are independent and that the common distribution is

$$p(x_i) = \phi\, e^{-\phi x_i} \qquad x_i > 0 \tag{3}$$

The observed stochastic process is thus that process consisting of the first, second, third ... events (fault discoveries). The times of these events are the order statistics $X_{(1)}$, $X_{(2)}$, $X_{(3)}$, ... The inter-event times are the spacings between the order statistics:

$$T_i = X_{(i)} - X_{(i-1)} \tag{4}$$

for $i = 2, 3, \ldots$ with $T_1 = X_{(1)}$. It is easy to show that this formulation via order statistics agrees with (1) and (2).

The unknown parameters of the model, N and ϕ, are estimated by maximum likelihood. This forms stage (ii) of the prediction system. Predictions are made (stage (iii)) by the 'plug-in' rule: substitution of these ML estimates into the appropriate model expressions. Thus, for example, when t_1, t_2, ..., t_{i-1} are the observed data, the predicted (current) reliability is

$$\tilde{R}_i(t) = e^{-(\hat{N}-i+1)\hat{\phi}t} \tag{5}$$

which is an estimate of

$$R_i(t) = P(T_i > t).$$

The most serious criticism of this model is that it assumes the debugging process is purely deterministic and that all faults contribute equally to the unreliability of a program. This seems to be very implausible, and recent empirical studies (Adams (1984), Nagel & Skrivan (1981)) suggest that in reality the sizes of ϕ's corresponding to different faults will differ by orders of magnitude. We shall show, in a later section, that the use of this model in a context where the ϕ's really are very variable may result in reliability predictions which are too optimistic.

3.2 Bayesian Jelinski-Moranda (BJM)

There has been considerable research into properties of the ML parameter estimates of JM (Forman & Singpurwalla (1977), Littlewood & Verrall (1981), Joe & Reid (1985)). This was motivated by the apparently poor predictions of the model in many cases, and the suspicion that this poor predictive capability might be the fault of stages (ii) and (iii), rather than stage (i), of the prediction system.

BJM (Littlewood & Sofer (1985)) is an attempt to remove the problems of using ML and a 'plug-in' rule for (ii) and (iii) by using a proper Bayesian predictive set-up. For reasons of mathematical tractability BJM is a slightly different model from JM, being parameterised as (λ, ϕ) with $\lambda \equiv N\phi$. Here λ can be regarded as the initial rate of occurrence of failures, and f the improvement resulting from a fix. The modelling difference is that λ is not constrained to be an integer multiple of ϕ .

The Bayesian analysis proceeds in the usual way: We start with a prior distribution for (λ, ϕ). In Keiller *et al* (1983 b) this is taken to be of independent gamma type; here we shall take the 'ignorance prior' member of this gamma family. Given data $t_1, ..., t_{i-1}$ the posterior distribution

$$p(\lambda,\phi f| \ t_1, ..., t_{i-1}) \tag{6}$$

can be computed in the usual way to form stage (ii) of the prediction system. Bayesian predictive distributions are used for stage (iii) (Aitchison & Dunsmore (1975)) so that, for example, the current reliability function is

$$\tilde{R}_i(t) = \iint R_i(t| \ \lambda, \ \phi)p(\lambda, \ \phi|t_1, ..., t_{i-1})d\lambda d\phi \tag{7}$$

where the conditional reliability is

$$R_i(t| \ \lambda, \ \phi) = e^{-(\lambda-[i-1]\phi)t} \tag{8}$$

Our experience with this prediction system is that it generally only offers marginally improved predictions compared with JM itself. This would suggest that the **model** itself, stage (i) of the prediction system, is at fault.

3.3. Littlewood (L)

This model (Littlewood (1981)) is an attempt to answer the criticisms of JM/BJM whilst retaining the finite fault-count, order statistic approach. The major drawback of JM/BJM is that it treats debugging as a deterministic process: each fix is effective with certainty and all fixes have the same effect on the reliability. In L, it is assumed that faults contribute different amounts to the unreliability of the software. Thus, although it is assumed that fixes are effective with certainty, the sequence of (improving) failure rates forms a stochastic process since the magnitude of each improvement is unpredictable.

In detail, the model assumes, as before

$$p(t_i|\Lambda_i = \lambda_i) = \lambda_i e^{-\lambda_i t_i} \tag{9}$$

where the random variables $\{\Lambda_i\}$ represent the successive ROCOF's arising from the gradual elimination of faults. Here

$$\Lambda_i = \Phi_1 + \Phi_2 + \ldots + \Phi_{N-i+1} \tag{10}$$

where N is the initial number of faults and Φ_j represents the (random variable) rate associated with fault j (in arbitrary labelling).

The initial rates $\Phi_1, \ldots, \Phi_N$ are assumed to be independent, identically distributed gamma(α,β) random variables. When the program has executed for a total time τ, use of Bayes theorem shows that the *remaining* rates are iid gamma($\alpha,\beta+\tau$) random variables. This reflects our intuitive belief that the early fixes will tend to be associated with faults having larger rates: the initial average fault size is α/β, which becomes $\alpha/(\beta + \tau)$ for the faults remaining at time τ. Execution depletes the program of larger faults.

As was the case for JM, this model can be interpreted via order statistics. If we let X_i represent the time to detect fault i (in our arbitrary labelling), then

$$p(x_i|\Phi_i = \phi_i) = \phi_i e^{-\phi_i x_i} \qquad x_i > 0 \tag{11}$$

and Φ_i is a gamma(α,β) variate. Thus unconditionally the pdf of x_i is Pareto (cf. Newby chapter 3 section 2.3 and chapter 4 section 8):

$$p(x_i) = \frac{\alpha\beta^{\alpha}}{(\beta + x_i)^{\alpha+1}} \qquad x_i > 0 \tag{12}$$

by mixing (11) over the gamma distribution of Φ_i. The observed stochastic process of inter-failure times T_1, T_2, ... is then the process of *spacings* of the order statistics of the iid Pareto X's:

$$\begin{aligned} T_1 &= X_{(1)} \\ T_i &= X_{(i)} - X_{(i-1)} \qquad i = 2, 3, \dots \end{aligned} \tag{13}$$

Estimation of the unknown parameter is by ML and prediction by substituting these into appropriate model expressions via the 'plug-in' rule. The estimated current reliability, based on data $t_1, t_2, \dots, t_{i-1}$, is then

$$\tilde{R}_i(t) = \left(\frac{\tilde{\beta} + t}{\tilde{\beta} + \tau + t}\right)^{(\hat{N}-i+1)\hat{\alpha}} \tag{14}$$

where

$$\tau = \sum_{j=1}^{i-1} t_j \tag{15}$$

is total elapsed time.

A proper Bayesian analysis of this model seems difficult, largely because of the role played by β in the likelihood function. We have considered briefly elsewhere (Abdel Ghaly (1986)) an *ad hoc* approach. This begins by assuming initially that β is known, whereupon it is possible to perform a conventional Bayesian analysis of the unknown (λ,ϕ), where $\lambda = N\alpha$. This uses independent gamma priors and the posterior analysis can be conducted analytically. For any given β it is then possible to form the usual predictive distributions. Finally an estimator of β is used based on a maximum likelihood approach.

3.4. Littlewood and Verrall (LV)

This model (Littlewood & Verrall (1973) again treats the successive rates of occurrence of failures, as fixes take place, as random variables. As in JM and BJM, it assumes

$$p(t_i|\Lambda_i = \lambda_i) = \lambda_i e^{-\lambda_i t_i} \qquad t_i > 0 \tag{16}$$

The sequence of rates Λ_i is treated as a sequence of independent *stochastically decreasing* random variables. This reflects the likelihood, but not certainty, that a fix will be effective. It is assumed that

$$p(\lambda_i) = \frac{[\psi(i)]^{\alpha} \lambda_i^{\alpha-1} e^{-\psi(i)\lambda_i}}{\Gamma(\alpha)} \tag{17}$$

a gamma distribution with parameters α, $\psi(i)$.

The function $\psi(i)$ determines the reliability growth. If, as is usually the case, $\psi(i)$ is an increasing function of i, it is easy to show that $\{\Lambda_i\}$ forms a stochastically decreasing sequence. Notice how this contrasts with the JM/BJM case where fixes are *certain* (and of equal magnitude). For LV a fix *may* make the program less reliable, and even if a improvement takes place it is of uncertain magnitude.

The choice of parametric family for $\psi(i)$ is under the control of the user. In this paper we shall take

$$\psi(i) = \psi(i,\beta) = \beta_1 + \beta_2 i \tag{18}$$

Predictions are made by ML estimation of the unknown parameters α, β_1, β_2 and use of the 'plug-in' rule. Thus the estimate of the current reliability function after seeing inter-failure times $t_1, t_2, \ldots, t_{i-1}$ is

$$\tilde{R}_i(t) = \left[\frac{\psi(i,\hat{\beta})}{t + \psi(i,\hat{\beta})}\right]^{\hat{\alpha}} \tag{19}$$

where α, β are the ML estimates of the parameters.

3.5. Keiller and Littlewood (KL)

KL (Keiller *et al* (1983 a and b) is similar to LV, except that reliability growth is induced via the *shape* parameter of the gamma distribution for the rates. That is, it makes assumption (14) with

$$p(\lambda_i) = \frac{\beta^{\psi(i)} \lambda_i^{\psi(i)-1} e^{-\beta\lambda_i}}{\Gamma(\psi(i))} \tag{20}$$

Here reliability growth, represented by stochastically decreasing rates (and thus stochastically increasing T's), occurs when $\psi(i)$ is a *decreasing* function of i. Again, choice of the parametric form of $\psi(i)$ is under user control. Here we shall use

$$\psi(i,\alpha) = (\alpha_1 + \alpha_2 i)^{-1} \tag{21}$$

Prediction is again by ML estimation and the 'plug-in' rule, so that, for example, the estimated current reliability function after observing t_1, $t_2, \ldots, t_{i-1}$, is

$$\tilde{R}_i(t) = \left[\frac{\hat{\beta}}{t + \hat{\beta}} \right]^{\psi(i,\hat{\alpha})} \tag{22}$$

3.6. Weibull order statistics (W)

The JM and L models can be seen as particular examples of a general class of stochastic processes based on order statistics. These processes exhibit inter-event times which are the spacings between order statistics from a random sample of N observations with pdf f(x). For JM and L, f(x) is respectively exponential and Pareto. For the W model we assume f(x) is the Weibull density

$$f(x) = \alpha\beta\, x^{\beta-1} e^{-\alpha x^{\beta}} \tag{23}$$

Estimation from realisations of the T_i random variables is via ML, and prediction via the 'plug-in' rule. Details of this model are published elsewhere (Abdel Ghaly (1986)).

Other models from this general class of stochastic processes seem attractive candidates for further study (Miller (1986)).

3.7 Duane (D)

The Duane model originated in hardware reliability studies. Duane (1964) claimed to have observed in several disparate applications that the reliability growth in hardware systems showed the ROCOF having a power law form in operating time. Crow (1977) took this observation and added the assumption that the failure process was a non-homogeneous Poisson process (NHPP) with rate

$$kbt^{b-1} \qquad (k,\ b,\ t > 0) \tag{24}$$

There is a sense in which an NHPP is inappropriate for software reliability growth. We know that it is the fixes which change the reliability, and these occur at a finite number of known times. The true rate presumably changes discontinuously at these fixes, whereas the NHPP rate changes continuously. However, it is known (Miller (1986) that for a single realisation it is not possible to distinguish between an order statistic model and an NHPP with appropriate rate.

Prediction from this model involves ML estimation and the 'plug-in' rule.

3.8. Goel-Okumoto (GO)

It is easy to show (Miller (1986)) that if we treat the parameter N in the JM model as a Poisson random variable with mean m, the unconditional process is exactly an NHPP with rate function

$$m\ \phi\ e^{-\phi t} \tag{25}$$

Presumably such a mixture over a distribution for N only makes sense to a subjective Bayesian, for which this distribution could be taken to represent 'his' uncertainty about N.

Prediction for this model is, again, via ML estimation and the 'plug-in' rule.

Details can be found elsewhere (Goel & Okumoto (1979)).

Miller (1986), in an interesting recent paper, shows that this NHPP and the JM model are indistinguishable on the basis of a single realisation of the stochastic process, t_1, t_2,... . He notes, however, that inferences for the two would differ, since they have different likelihood functions. This implies that *predictions* based on ML inference and the 'plug-in' rule would be different.

This is a very curious situation. We have two different prediction systems, giving different predictions, but based upon *models* which are indistinguishable on the basis of the data.

3.9. Littlewood NHPP (LNHPP)

This model is an NHPP with rate function

$$\frac{m \alpha \beta^{\alpha}}{(\beta + t)^{\alpha+1}}$$

Again this can be interpreted as the Littlewood model mixed over a Poisson distributed N variable.

Prediction is via ML estimation and the 'plug-in' rule.

Similar indistinguishability conditions exist between L and LNHPP as were considered in section 3.9.

4. Examples of Use

The simplest question a user can ask is: how reliable is my program now? As debugging proceeds, and more inter-failure time data is collected, this question is likely to be repeated. It is hoped that the succession of reliability estimates will show a steady improvement due to the removal of faults.

Let us assume that our user is a simple man, and will be satisfied with an accurate estimate of the *median time to next failure* at each stage. He decides

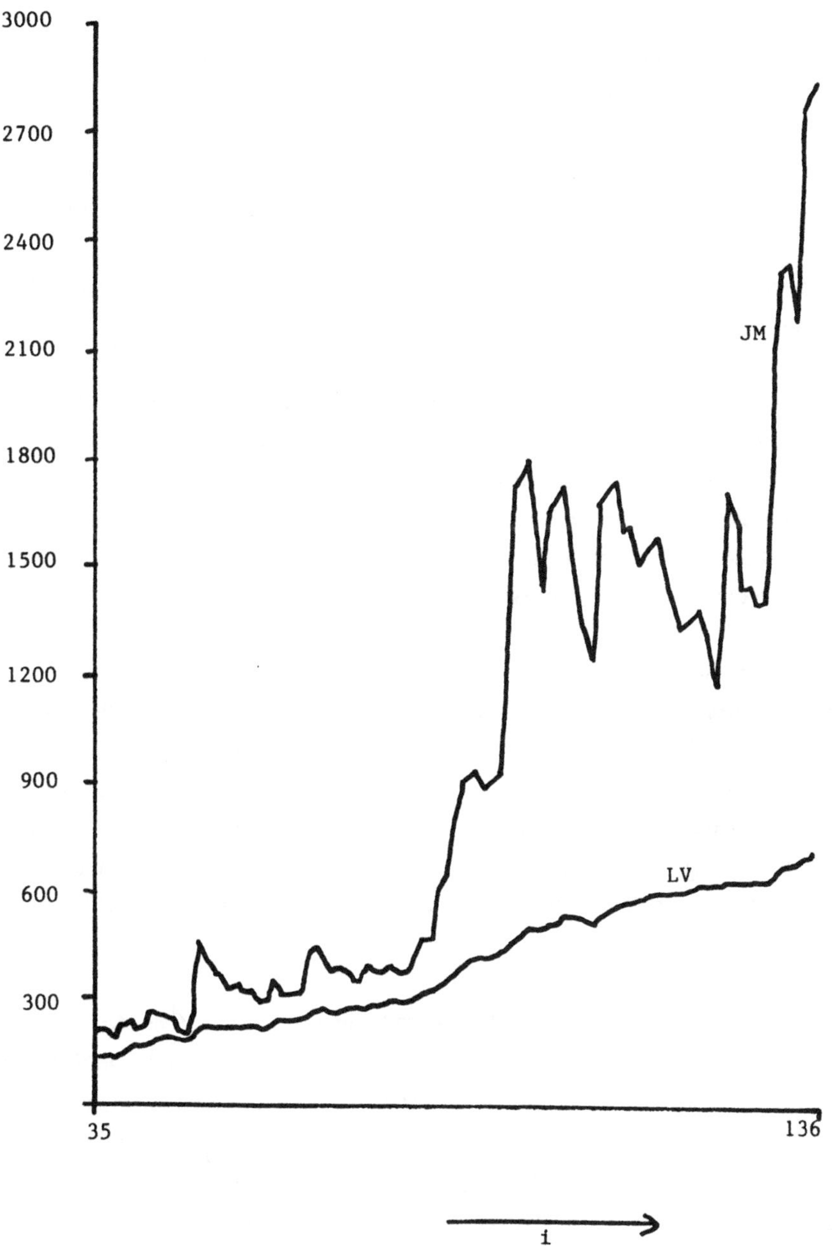

Figure 3. Median plots for JM and LV, data from Table 1. Plotted are predicted median of T_i (based on $t_1, t_2, \ldots, t_{i-1}$) against i.

that he will make his calculations using both JM and LV. Figure 3 shows the results he would get if he were using the data shown in Table 1. At each stage, for a particular prediction system, the point plotted is the predicted median of the time to next failure, T_i, based on the available data t_1, t_2, ..., t_{i-1}. Such plots are thus a simple way of tracking progress in terms of estimated achieved reliability.

Our user would, we believe, be alarmed at the results. Whilst the models agree that reliability growth is present, they disagree profoundly about the nature and extent of that growth. The JM predictions, particularly at later stages, are much more optimistic than those of LV: JM suggests that the reliability is greater than LV would suggest.

In addition, the JM predictions are more 'noisy' than those of LV. The latter suggests that there is a steady reliability growth with no significant reversals. JM suggests that important set-backs are occurring. What should the user do? He might be tempted to try yet more prediction systems, and hope to arrive somehow at consensus. If he were to adopt this approach, the chances are that his confusion would increase.

The important point is that he has no guide as to which, if any, of the available predictions is close to the truth. Is the true reliability as high as suggested by JM in Figure 3? Or are the more conservative estimates of LV nearer to reality? Perhaps as important: are the apparent decreases in reliability indicated by JM real (bad fixes?), or artifacts of the statistical procedures? (The JM model does not, in fact, allow for the possibility of bad fixes, so these reversals must be due to stages (ii) and (iii) of the prediction system.)

If our user wished to predict further ahead than the next time to failure, he would find the picture even bleaker. Figure 4 shows how JM and LV perform when required to predict a median 20 steps ahead. That is, prediction is made of the median of T_i, at stage i–20, using observations t_1, t_2,..., t_{i-20}. It is obvious in this case that JM is performing very badly. Its excursions to infinity are caused by its tendency to suggest that at stage i–20 there are less than 20 faults left in the program, so that the estimated median of T_i is infinite. At least LV does not behave in this absurd fashion. In addition it seems reasonably 'self-consistent' in that the prediction of the median of T_i made 20 steps before (based on t_1, ..., t_{i-20}) is usually in good agreement

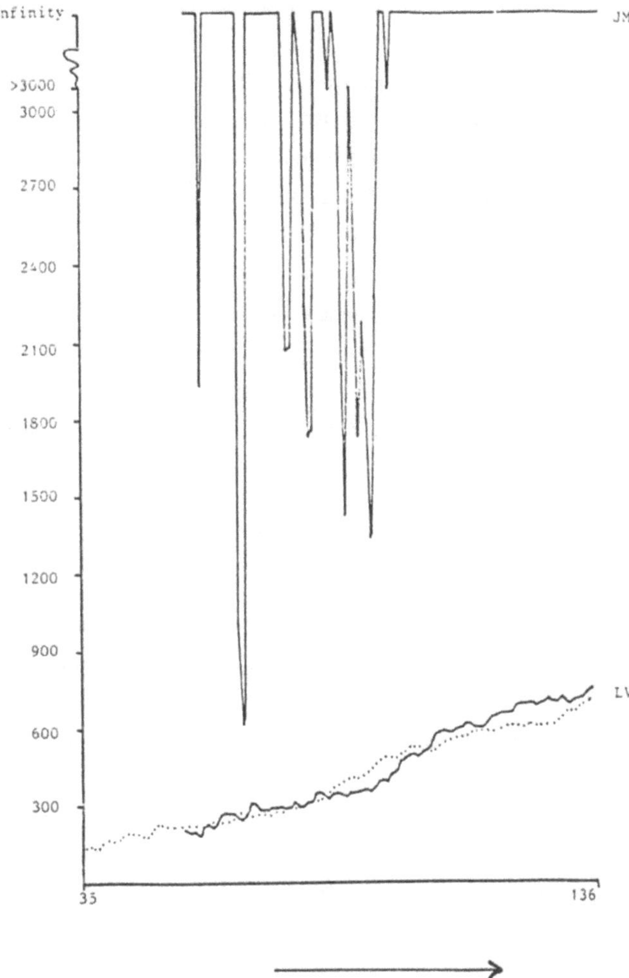

Figure 4. Median predictions 20 steps ahead for JM, LV using data of Table 1. JM makes many excursions to infinity, because it frequently estimates the number of remaining faults to be less than 20! The LV predictions 20 steps ahead are in close agreement with the (later) 1 step ahead median prediction (shown dotted). This is a useful 'self-consistency' property: a prediction system ought to have the property that a prediction from T_i based on t_i, ..., t_{i-20} is 'close' to a later one based on t_i, ..., t_{i-j}. Clearly JM does *not* have this property: compare with Figure 3.

with the later 'current median' estimate (based on t_i, t_{i-1}, ...). Even such self-consistency, though, is no guarantee that these predictions are close *to the truth.*

These kinds of disagreement between different solutions to the prediction problem are very common. Until recently users had no way of deciding which, if any, reliability metrics could be trusted. All that was available was a great deal of special pleading from advocates of particular models: 'trust my model and you can trust the predictions'. This is not good enough. No model is totally convincing on *a priori* grounds. More importantly, a 'good' model is only one of the three components needed for good predictions.

In the next section we describe some ways in which a user of reliability models can obtain insight into their performance *on his/her own data.*

5. Analysis of Predictive Quality

We shall concentrate, for convenience, upon the simplest prediction of all concerning *current reliability*. Most of these techniques can be adapted easily to some problems of longer-term prediction, but there are also novel difficulties arising from these problems. We shall return to this question later.

Having observed t_1, t_2, ..., t_{i-1} we want to predict the random variable T_i. More precisely, we want a good estimate of

$$F_i(t) = P(T_i < t) \tag{27}$$

or, equivalently, of the reliability function

$$R_i(t) = 1 - F_i(t) \tag{28}$$

From one of the prediction systems described earlier we can calculate a predictor

$$\tilde{F}_i(t) \tag{29}$$

A user is interested in the 'closeness' of $\tilde{F}_i(t)$ to the unknown *true* $F_i(t)$. In fact, he/she may be only interested in summary statistics such as mean (or median) time to failure, ROCOF, etc. However, the quality of these summarised predictions will depend upon the quality of $\tilde{F}_i(t)$, so we shall concentrate on the latter.

Clearly, the difficulty of analysing the closeness of $\tilde{F}_i(t)$ to $F_i(t)$ arises from our never knowing, even at a later stage, the true $F_i(t)$. If this were available (for example, if we simulated the reliability growth data from a sequence of known distributions) it would be possible to use measures of closeness based upon *entropy* and *information* (Akaike (1982)), or distance measures such as those due to Kolmogorov or Cramer–von Mises (Kendall & Stuart (1961)).

In fact, the only information we shall obtain will be a single realisation of the random variable T_i when the software next fails. That is, after making the prediction $\tilde{F}_i(t)$ based upon t_1, t_2, ..., t_{i-1}, we shall eventually observe t_i, which is a sample of size one from the true distribution $F_i(t)$. We must base all our analysis of the quality of the predictions upon these pairs $\{\tilde{F}_i(t), t_i\}$.

Our method will be an emulation of how a user would informally respond to a sequence of predictions and outcomes. He/she would inspect the pairs $\{\tilde{F}_i(t), t_i\}$ to see whether there is any evidence to suggest that the t_i's are not realisations of random variables from the $F_i(t)$'s. If such evidence were found, it would suggest that there are significant differences between $\tilde{F}_i(t)$ and $F_i(t)$, i.e. that the predictions are not in accord with actual behaviour. The 20–step ahead predictions of JM shown in Figure 4 are an example of strong evidence of disagreement between prediction and outcome: the predictions are often of infinite time to failure (program fault–free, so $F_i(t) = 0$ for all t), but the program always fails in finite time.

Consider the following sequence of transformations:

$$u_i = \tilde{F}_i(t_i) \tag{30}$$

Each is a probability integral transform of the observed t_i, using the previously calculated predictor $\tilde{F}_i$ based upon t_1, t_2, ..., t_{i-1}. Now, if each $\tilde{F}_i$ were identical to the true F_i, it is easy to see that the u_i would be realisations of independent uniform U(0,1) random variables (Rosenblatt (1952), Dawid (1984 b)). Consequently we can reduce the problem of examining the closeness of $\tilde{F}_i$ to F_i (for some range of values of i) to the question of whether the sequence $\{u_i\}$ 'looks like' a random sample from U(0,1). Readers interested in the more formal statistical aspects of these issues should consult the recent work of Dawid (1984 a and 1984 b).

We consider now some ways in which the $\{u_i\}$ sequence can be examined.

5.1 The u-plot

Since the u_i's should look like a random sample from U(0,1) if the prediction system is working well, the first thing to examine is whether they appear *uniformly* distributed. We do this by plotting the sample cumulant distribution function (cdf) of the u_i's and comparing it with the cdf of U(0,1), which is

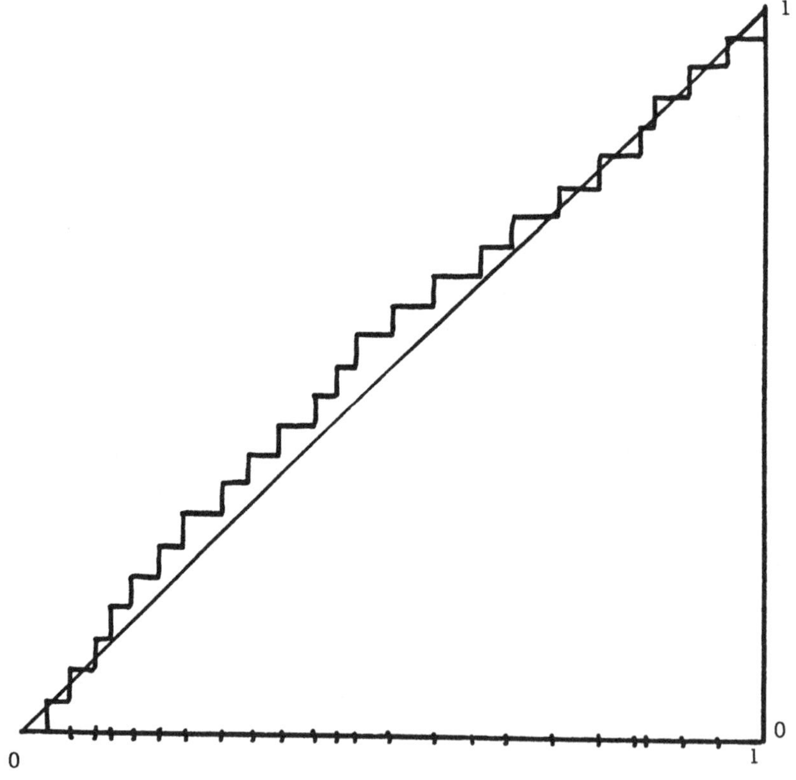

Figure 5. How to draw a u-plot. Each of the n u_i's, with a value between 0 and 1, is placed on the horizontal axis. The step function increased by 1/(n+1) at each of these points.

the line of unit slope through the origin. Figure 5 shows how such a u-plot is drawn. The 'distance' between them can be summarised in various ways. We shall use the Kolmogorov distance, which is the maximum absolute vertical difference

In Figure 6 are shown the u-plots for LV and JM predictions for the data of Table 1. The predictions here are $\tilde{F}_{36}(t)$ through $\tilde{F}_{135}(t)$. The Kolmogorov distances are 0.190 (JM) and 0.144 (LV). In tables of the Kolmogorov distribution the JM result is significant at the 1% level, LV only at the 5% level.

From this analysis it appears that neither set of predictions is very good, but that JM is significantly worse than LV.

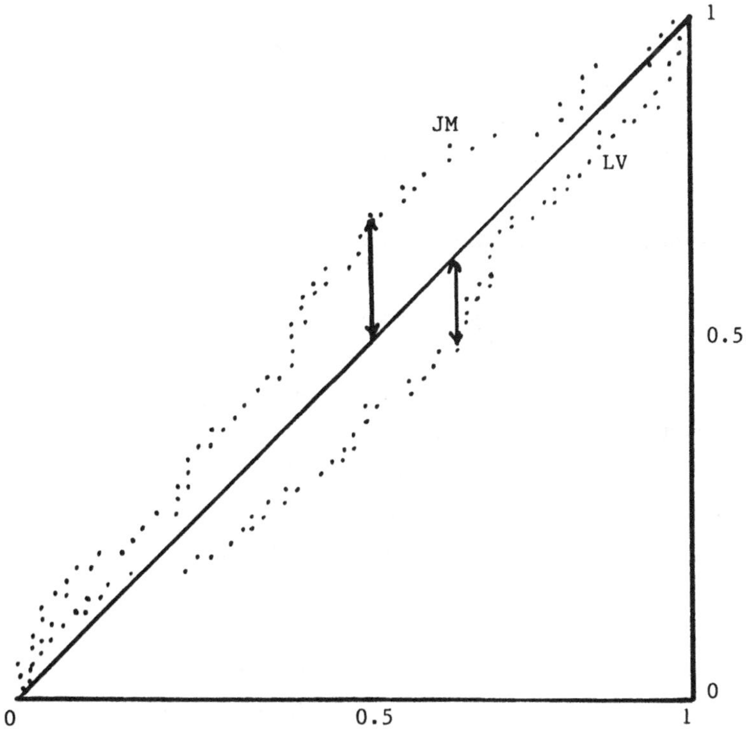

Figure 6. LV, JM u-plots, data of Table 1. Steps omitted for clarity. Note that these are reproduced from lineprinter plots and do not correspond exactly to true plot.

In fact the detailed plots tell us more than this. The JM plot is everywhere *above* the line of unit slope (the U(0,1) cdf); the LV plot almost everywhere below it. This means that the u_i's from JM tend to be too small and those from LV too large. But u_i represents the predicted probability that T_i will be less than t_i, so consistently too small u_i's suggest that the predictions are underestimating the chance of small t's. That is, the JM plot tells us that these predictions are too *optimistic*; the LV plot that these predictions are too *pessimistic* (although to a less pronounced degree).

There is evidence from this simple analysis, then, that the truth might lie somewhere between the predictions from JM and LV, but probably closer to LV. In particular, the true median plot probably lies between the two plots of Figure 3. We shall return to this idea in a later section, where further evidence will be given for our belief that the two prediction systems bound the truth for this data set.

At this stage, then, a user might take an analysis of this kind to help him make further predictions. He might, for example, adopt a conservative position and decide to use LV for his next prediction, and be reasonably confident that he would not over estimate the reliability of the product.

5.2. The y-plot, and scatter plot of u's

The u-plot treats one type of departure of the predictors from reality. There are other departures which cannot be detected by the u-plot. For example, in one of our investigations we found a data set for which a particular prediction system had the property of optimism in the early predictions and pessimism in the later predictions. These deviations were averaged out in the u-plot, in which the temporal ordering of the u_i's disappears, so that a small Kolmogorov distance was observed. It is necessary, then, to examine the u_i's for *trend*.

Figure 7 shows one way in which this can be done. First of all, it should be obvious that, since each u_i is defined on (0,1), the sequence u_i (Stage 1 in Figure 7) will look super-regular. The transformation $x_i = -\ln(1-u_i)$ will produce a realisation of iid unit exponential random variables if the $\{u_i\}$ sequence really are a realisation of iid U(0,1) random variables. That is, Stage 2 of Figure 7 should look like a realisation of a homogeneous Poisson process; the alternative hypothesis (that there is trend in the u_i's) will

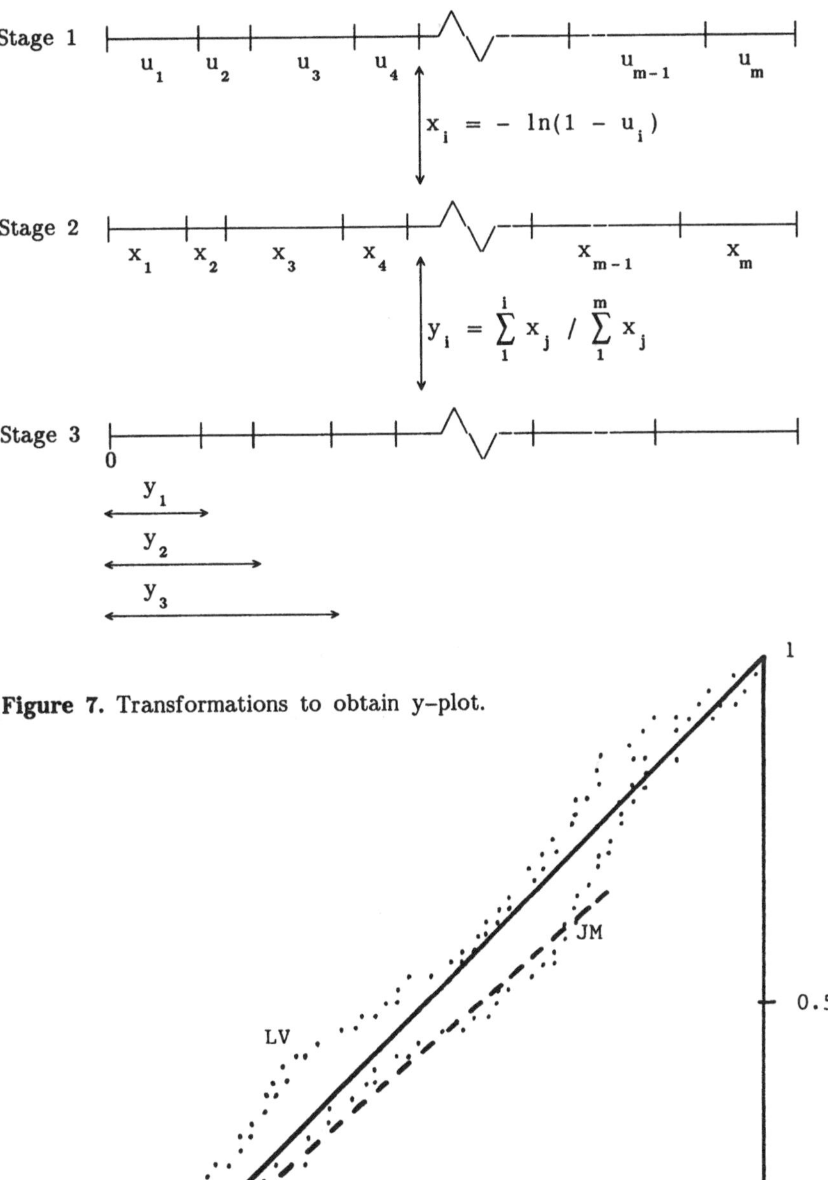

Figure 7. Transformations to obtain y–plot.

Figure 8. JM and LV y–plots for data of Table 1. Again, these are line–printer plots and points do not correspond exactly to true points.

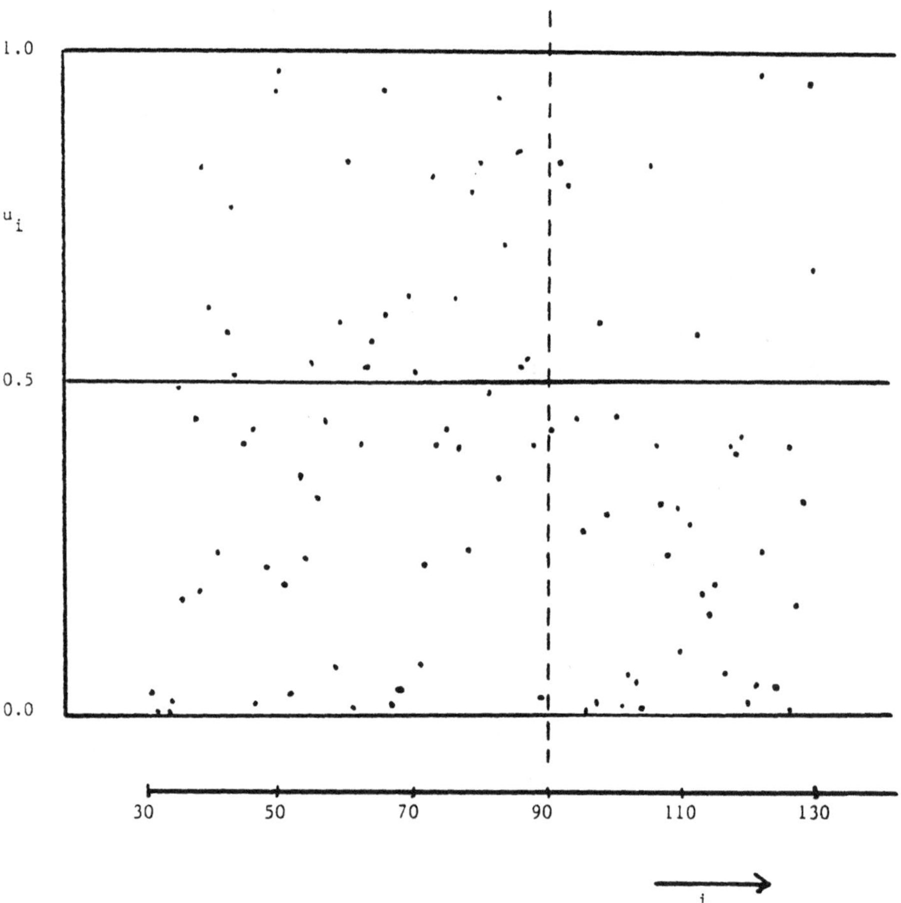

Figure 9. Scatter plot of u_i against i for JM predictions from Table 1 data. There are 'too many' small u's to the right of the dotted line.

show itself as a non–constant rate for this process. One simple test is to normalise the Stage 2 process onto (0,1), as in Stage 3 of Figure 7, and plot as in the previous section (Cox & Lewis (1966)).

Other procedures could be adopted, for example the Laplace test (Ascher & Feingold (1984), Cox & Lewis (1966), but we think that the plots are more informative. For example, see Figure 8 where this *y-plot* procedure is applied to the LV and JM predictions of the Table 1 data. The Kolmogorov distances are 0.120 (JM) and 0.110 (LV), neither of which are significant at the 10% level. More interestingly, a close examination of the JM y–plot suggests that it is very close to linearity in the early stages (until about i = 90: see broken line).

This observation is confirmed by a *scatter plot* of u_i against i: Figure 9. After i = 90 there are only 8 out of 39 u_i's greater than 0.5.

The implication is that the too optimistic predictions from JM are occurring mainly after i = 90. That is, the poor performance arises chiefly from the later predictions. Since these are based upon larger amounts of data, it is unlikely that stages (ii) and (iii) of the prediction system are responsible.

The effect can be seen quite clearly in the median plots (Figure 3). We can now have reasonable confidence that the sudden increase in the median plot of JM, at about i = 90, is not a true reflection of reliability of the software under study. It is noticeable that this effect does not occur in the LV predictions.

5.3. Measures of 'noise'

It is instructive at this stage to digress a little and consider briefly the estimation problem in classical statistics. There we have a random sample (independent, *identically distributed* random variables) from a population with an unknown parameter, θ. If we assume, for simplicity, that θ is scalar, it is usual to seek an estimator for θ, say $\tilde{\theta}$, which has small *mean square error*:

$$\text{mse}(\theta) = E\{(\tilde{\theta} - \theta)^2\} \tag{31}$$

$$= \text{Var}(\tilde{\theta}) + (\text{bias } \tilde{\theta})^2 \tag{32}$$

There is thus a trade-off between the variance of the estimator and its bias. It is not obvious, without adopting extra criteria, how one would choose among estimators with the same mse but different variances and biases.

In our prediction problem the situation is much more complicated: we wish at each stage to estimate a function, $F_i(t)$, not merely a scalar; and the context is non-stationary since the $F_i(t)$'s are changing with i.

However, the analogy with the classical case has some value. We can think of the u-plot as similar to an investigation of bias. Indeed, it is easy to show that, if $E\{\tilde{F}_i(t)\} = F_i(t)$ for all i, the expected value of the u-plot is the line of unit slope. Thus a systematic deviation between $E\{\tilde{F}_i(t)\}$ and $F_i(t)$ will be detected by the u-plot. We shall return to this question when we look at adaptive procedures in a later section.

The fact that we are making a sequence of predictions in a non-stationary context complicates matters. Thus a prediction system could be biased in one direction for early predictions and in the other direction for later predictions (and, of course, more complicated deviations from reality are possible). The y-plot is a (crude) attempt to detect such a situation.

The u-plot and y-plot procedures then, are an attempt to analyse something analogous to bias. Can we similarly analyse 'variability' in our more complicated situation?

The median plot of Figure 3, for example, shows JM to be more variable than LV. This suggests that the $\{\tilde{F}_i(t)\}$ sequence for JM is more variable than that for LV. The important question is whether this extra variability of JM is an accurate reflection of what happens to the true $\{F_i(t)\}$. Is $\{\tilde{F}_i(t)\}$ fluctuating rapidly in order to track the truly fluctuating $\{F_i(t)\}$, or is it exhibiting random sampling fluctuations about a slowly changing $\{F_i(t)\}$ sequence?

If we had the true $\{F_i(t)\}$ sequence available, it would be relatively easy to obtain measures akin to variance. We could, for example, average the Cramer-von Mises distances between $\tilde{F}_i(t)$ and $F_i(t)$ over some range of i. Unfortunately, the $\{F_i(t)\}$ sequence is not known, and we have been unsuccessful in our attempts to obtain good measures of the variability between $\{\tilde{F}_i(t)\}$ and $\{F_i(t)\}$. There follow some quite crude measures of variability. In section 5.4 we shall consider a global measure which incorporates both 'bias' and 'noise': loosely analogous to mse in the iid case.

5.3.1. Braun statistic

Braun has proposed (Braun & Paine (1977)), on obvious intuitive ground, the statistic

$$\frac{\sum_i \{t_i - \tilde{E}(T_i)\}^2}{\sum_i \{t_i - \bar{t}\}^2} \frac{n-1}{n-2} \tag{33}$$

where $\tilde{E}(T_i)$ is the estimated mean of T_i, i.e. the expectation of the predictor distribution, $\tilde{F}(T_i)$, and n is the number of terms in the sums. The normalising denominator is not strictly necessary here, since it will the be the same for all prediction systems and we shall only be *comparing* values of this statistic for different systems on the same data: there are no obvious ways of carrying out formal tests to see whether a particular realisation of the statistic is 'too large'.

5.3.2 Median variability

A comparison of

$$\sum_i \left| \frac{m_i - m_{i-1}}{m_{i-1}} \right| \tag{34}$$

where m_i is the predicted median of T_i, between different prediction systems can indicate objectively which is producing the most variable predictions. For example, the greater variability of the JM medians in Figure 3 is indicated by a value of 9.57 against LV's 2.96. Of course, this does not tell us whether the extra JM variability reflects true variability of the actual reliability.

5.3.3 Rate variability

A similar comparison can be based on the ROCOF sequence, r_i, calculated immediately after a fix:

$$\sum_i \left| \frac{r_i - r_{i-1}}{r_{i-1}} \right| \tag{35}$$

The JM value for the predictions of Table 1 data is 8.37, for LV 3.18.

For both (34) and (35) we can only compare prediction systems on the same data. More importantly we cannot know whether the greater noisiness of a particular prediction system is "unjustified".

5.4 Prequential likelihood

In a series of important recent papers (1982, 1984 a, 1984 b, 1989), A. P. Dawid has treated theoretical issues concerned with the validity of forecasting systems. Dawid's discussion of the notion of *calibration* is relevant to the software reliability prediction problem. Here we shall confine ourselves to the *prequential likelihood* (PL) function and, in particular, the *prequential likelihood ratio* (PLR). We shall use PLR as an investigative tool to decide on the relative plausibility of the predictions emanating from two models.

The PL is defined as follows. The predictive distribution $\tilde{F}(t_i)$ for T_i based on $t_1, t_2, ..., t_{i-1}$ will be assumed to have a pdf

$$f_i(t) = F_i'(t) \tag{36}$$

For predictions of $T_{j+1}, T_{j+2}, ..., T_{j+n}$, the *prequential likelihood* is

$$PL_n = \prod_{i=j+1}^{j+n} F_i(t_i) \tag{37}$$

A comparison of two prediction systems, A and B, can be made via their *prequential likelihood ratio*

$$PLR_n = \frac{\prod_{i=j+1}^{j+n} \tilde{f}_i^A(t_i)}{\prod_{i=j+1}^{j+n} \tilde{f}_i^B(t_i)} \tag{38}$$

Dawid (1984 b) shows that if $PLR_n \rightarrow \infty$ as $n \rightarrow \infty$, prediction system B is discredited in favour of A.

To get an intuitive feel for the behaviour of the prequential likelihood, consider Figure 10. Here we consider for simplicity the problem of predicting a sequence of identically distributed random variables, i.e. $F_i(t) = F(t)$, $f_i(t) = f(t)$ for all i. The extension to our non-stationary case is trivial.

In Figure 10 the sequence of predictor densities are 'biased' to the left of the true distribution. Observations, which will tend to fall in the body of the true distribution, will tend to be in the (right hand) tails of the predictor densities. Thus the prequential likelihood will tend to be small.

In Figure 10 (b) the predictions are very 'noisy', but have an expectation close to the true distribution (low 'bias'). There is a tendency, again, for the body of the true distribution to correspond to a tail of the predictor (here either left or right tail). Thus the likely observations (from the body of the true distribution) will have low predictive probability, and the prequential likelihood will tend to be small again.

Notice that this last argument extends to our non-stationary case. Consider the case where the true distributions fluctuate for different values of i, corresponding to occasional bad fixes, for example. If the predictor sequence were 'too smooth', perhaps as a result of smoothing from an inference procedure, this would be detected. The observations would tend to fall in the bodies of the (noisy) true distributions, and hence in the tails of the predictors, giving a small prequential likelihood.

Thus the prequential likelihood can in principle detect predictors which are too noisy (when the true distributions are not variable) and predictors which are too smooth (when the true distributions are variable). This contrasts with the measures of variability proposed in section 5.3: here we could detect noise in a predictor, but could not tell whether it reflected actual noise in the reliability.

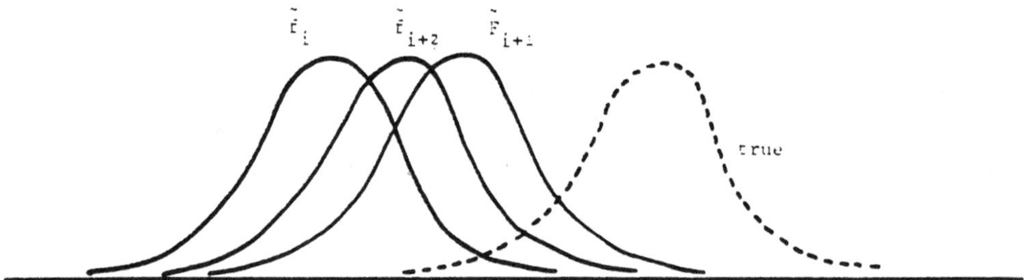

Figure 10 a. These predictions have high 'bias' and low 'noise'.

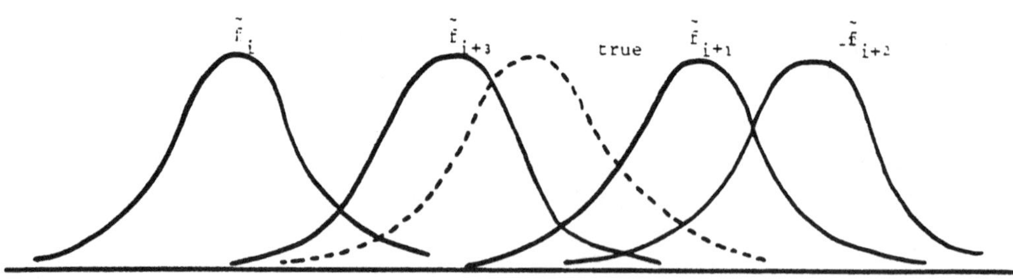

Figure 10 (b). These predictions have high 'noise' and low 'bias'.

The prequential likelihood, then, should allow us to detect both consistent deviations between prediction and reality ('bias'), and large variability in the distance between prediction and reality ('noise'). In this sense it is analogous to mse in parameter estimation (Miller (1983), Braun & Paine (1977)).

In fact it is possible to construct predictors which are (almost) exactly unbiased, but are useless in practice because of their great noisiness. An example is suggested by Miller (1983). He proposed an estimator based only upon the previous one or two observations. His idea was to assume that the $\{T_i\}$ sequence was of exponential random variables, and estimate the mean of T_{i-1} by using t_{i-1} or by using $(t_{i-1} + t_{i-2})/2$. In each case he contrived to obtain a predictor for T_{i-1}, $\tilde{F}_{i-1}(t)$, which was unbiased:

$$E\{\tilde{F}_{i-1}(t)\} = F_i(t) \tag{39}$$

An extra assumption, that $\tilde{F}_i(t)$ was close to $F_{i-1}(t)$, allowed the latter to be used as an approximate predictor for the *unobserved* T_1. Miller's intention was to produce an estimator which had a good u-plot ('unbiased') but which was clearly useless. A measure of his success can be seen by calculating the u-plot and y-plot Kolmogorov distance for his predictor (based on the previous two observations) on the data of Table 1. These are 0.078 and 0.069 respectively, which are not significant at the 10% level. These are much better than LV (0.14, 0.11) and JM (0.19, 0.12).

Could the prequential likelihood detect the (incorrect) noisiness of such a prediction system? Table 4 gives the PLR for JM versus Miller and LV versus Miller. In both cases we think it is obvious that the Miller predictions are being discredited ($PLR_n \rightarrow \infty$).

n	JM PLR	LV PLR
10	4.00	3.26
20	30.8	82.1
30	158	517
40	8.92×10^{4}	7.18×10^{5}
50	9.32×10^{5}	1.01×10^{6}
60	4.91×10^{6}	5.72×10^{5}
70	2.48×10^{6}	2.53×10^{7}
80	6.01×10^{5}	2.63×10^{8}
90	3.67×10^{6}	3.37×10^{10}
100	6.34×10^{8}	3.96×10^{11}

Table 4. This table shows the ability of PLR to reject an unbiased model which is *very* noisy. The Miller model predicts using only the last two observations. Here we show PLR values, at 10–step intervals, for JM versus Miller and LV versus Miller. Clearly, Miller is being discredited by each of the other prediction systems: even JM, which is known to be bad.

Of course, this is not a stringent test of the usefulness of PLR for discriminating between realistic good and bad prediction systems. In Table 5 is shown the PLR of LV against JM. There is evidence to reject JM in favour of LV. More importantly, there is again evidence that JM is doing particularly badly from about i = 95 (n =60) onwards. Prior to this, the relative fortunes of the two prediction systems fluctuate and it is briefly the case that JM is preferred.

n	PLR
10	1.19
20	0.318
30	0.252
40	0.096
50	0.745
60	6.50
70	0.088
80	0.00177
90	0.0000813
100	0.00119

Table 5. PLR of JM versus LV, data of Table 1. LV appears to discredit JM overall all, but less obvious for earlier predictions.

One interpretation of the PLR, when A and B are *Bayesian prediction systems* (Aitchinson & Dunsmore (1975)), is as an approximation to the posterior odds of model A against model B. Suppose that the user believes that either model A is true, with prior probability p(A), or model B is true with prior probability p(B) (= 1 − p(A)). He now observes the failure behaviour of the system; in particular, he makes predictions from the two prediction systems and compares them with actual behaviour via the PLR. Thus, when he had made predictions for T_{j+1}, T_{j+2}, ...,T_{j+n}, the PLR is

$$PLR_n = \frac{\prod_{i=j+1}^{j+n} \tilde{f}_i^A(t_i)}{\prod_{i=j+1}^{j+n} \tilde{f}_i^B(t_i)}$$

$$= \frac{p(t_{j+n}, \ldots, t_{j+1} | t_j, \ldots, t_1, A)}{p(t_{j+n}, \ldots, t_{j+1} | t_j, \ldots, t_1, B)} \qquad (40)$$

in an obvious notation. Using Bayes' Theorem this is

$$\frac{\dfrac{p(A|t_{j+n}, \ldots, t_1)\, p(t_{j+n}, \ldots, t_{j+1}|t_j, \ldots, t_1)}{p(A|t_j, \ldots, t_1)}}{\dfrac{p(B|t_{j+n}, \ldots, t_1)\, p(t_{j+n}, \ldots, t_{j+1}|t_j, \ldots, t_1)}{p(B|t_j, \ldots, t_1)}}$$

$$= \frac{p(A|t_{j+n}, \ldots, t_1)}{p(B|t_{j+n}, \ldots, t_1)} \cdot \frac{p(B|t_j, \ldots, t_1)}{p(A|t_j, \ldots, t_1)} \tag{41}$$

If the initial predictions were based only on prior belief ($j = 0$), the second term in (41) is merely the *prior odds ratio.* If the user is indifferent between A and B at this stage, this takes the value 1 since $p(A) = p(B) = 1/2$. Thus (41) becomes

$$\frac{w_A}{1 - w_A} \tag{42}$$

the posterior odds ratio, with w_A representing his posterior belief that A is true after seeing the data (i.e. after making predictions and comparing them with actual outcomes).

Of course, the prediction systems considered in section 3 are not all Bayesian ones. It is more usual to estimate the parameters via ML and use the 'plug-in' rule for prediction. Dawid (1984 b), however, shows that this procedure and the Bayesian predictive approach are asymptotically equivalent.

It is, in addition, not usual to allow $j = 0$ in practice. Although Bayesians can predict via prior belief without data, non-Bayesians usually insist that predictions are based on actual evidence. In practice, though, the value of j may be quite small.

With these reservations, we do think that (42) can be used as an intuitive interpretation of PLR. We shall use this idea, with some caution, in later sections.

6. Examples of Predictive Analyses

In this section we shall use the devices of the previous section to analyse the predictive quality of several prediction systems on the three data sets of Tables 1, 2 and 3. We emphasise that our primary intention is *not* to pick a 'universally best' prediction system. Rather, we hope to show how a fairly informal analysis can help a user to select reliability predictions in which he/she can have trust. Our own analyses suggest that one should approach a new data source without preconceptions as to which prediction system is likely to be best: such preconceptions are likely to be wrong.

Consider first the data of Table 1. Table 6 summarises our results concerning the quality of performance of the various prediction systems on this data.

In Table 6, it can be seen that LNHPP comes first on the PL ranks, followed by L, then BL, LV, KL and W. The Braun statistic rankings closely follow the PL ranks.

Both L and LNHPP have non significant u–plot and y–plot distances, although LNHPP has a smaller u–plot distance. This might suggest that LNHPP is slightly better on the 'bias' criterion. For noise, each of these prediction systems has similar rankings on the median and rate statistics, but in each case the value for LNHPP is smaller than that for L.

In fact, the predictions from L and LNHPP are very similar. Figure 11(a) shows their median predictions along with those of JM and LV shown earlier in Figure 3. Notice that these predictions are less pessimistic than LV, less optimistic than JM, but are closer to LV: this adds weight to our analysis of the LV and JM predictions in section 5. The slightly worse PL value for L compared with LNHPP is probably accounted for by its extra noise, as shown by the median and rate difference statistics.

Of the high PL–ranking predictions, that leaves BL, KL and W. Unfortunately, it is not easy to calculate medians for BL, so these are not shown on Figure 11(b). KL gives results which are very similar to LV: significantly too pessimistic on the u–plot. W is non significant on both u– and y–plots but (as for L) is noisier on both the statistics and the plot. Since u– and y–plot distances are so small, this noisiness probably explains its relatively poor performance on the PL.

Model	$\Sigma \left\| \frac{m_i - m_{i-1}}{m_{i-1}} \right\|$ (rank)	$\Sigma \left\| \frac{r_i - r_{i-1}}{r_{i-1}} \right\|$ (rank)	Braun statistic (rank)	u-plot KS distance (sig. level)	y-plot KS distance (sig. level)	$-\log_e$ (PL), ranks
JM	9.57 (8)	8.37 (10)	1.31 (10)	.190 (1%)	.120 (NS)	770.3 (9)
BJM		7.21 (8)	1.11 (9)	.170 (1%)	.116 (NS)	770.7 (10)
L(1)	6.82 (5)	6.23 (6)	0.86 (3)	.109 (NS)	.069 (NS)	762.4 (2)
BL		3.72 (4)	0.83 (2)	.119 (NS)	.075 (NS)	763.0 (3)
LV	2.96 (2)	3.18 (2)	0.90 (7)	.144 (5%)	.110 (NS)	764.7 (6)
KL	2.79 (1)	3.26 (3)	0.88 (5)	.138 (5%)	.109 (NS)	764.7 (5)
D	3.11 (3)	2.92 (1)	0.89 (6)	.159 (2%)	.093 (NS)	765.3 (7)
GO	8.62 (7)	7.34 (9)	1.11 (8)	.153 (2%)	.125 (10%)	768.6 (8)
LNHPP	4.16 (4)	3.84 (5)	0.82 (1)	.081 (NS)	.064 (NS)	761.4 (1)
W	7.38 (6)	6.61 (7)	0.86 (4)	.075 (NS)	.075 (NS)	763.0 (3)

Table 6. Analysis of data of Table 1 using the prediction systems described in the paper. In this, and the following tables, the instantaneous mean time to failure (IMTTF) is used instead of the predicted mean in the calculating of the Braun statistic for BJM, BL, D, CO, LNHPP, W. This is because the mean does not exist, or is hard to calculate. The IMTTF is defined to be the reciprocal of the current ROCOF.

Note[(1)] that for L the ML routine does not always terminate normally, so we are not certain that true ML estimates are being used. It is possible that we could get better L predictors by allowing the ML search routine to run longer.

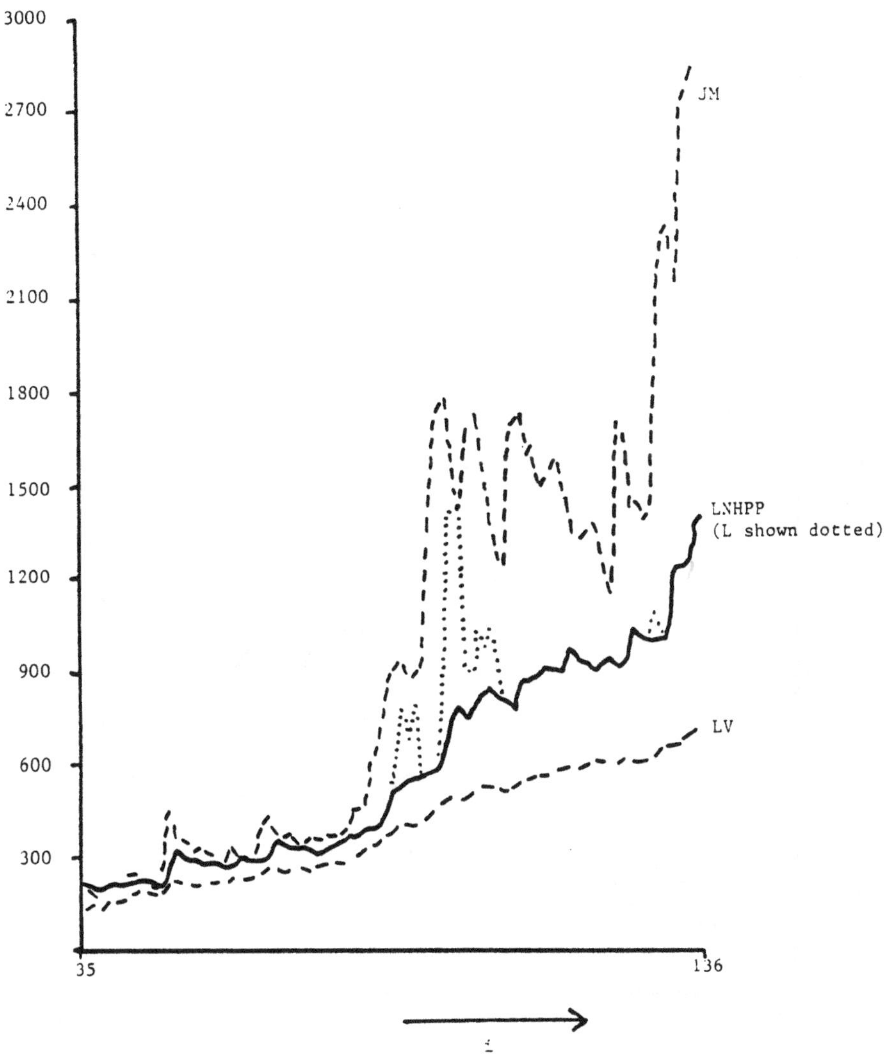

Figure 11 a. Median predictions from L, NHPP, LV, JM for data of Table 1. L and LNHPP are virtually indistinguishable for many predictions. The (dotted) excursions of L could be "spurious', resulting from non-convergence of the ML optimisation algorithm.

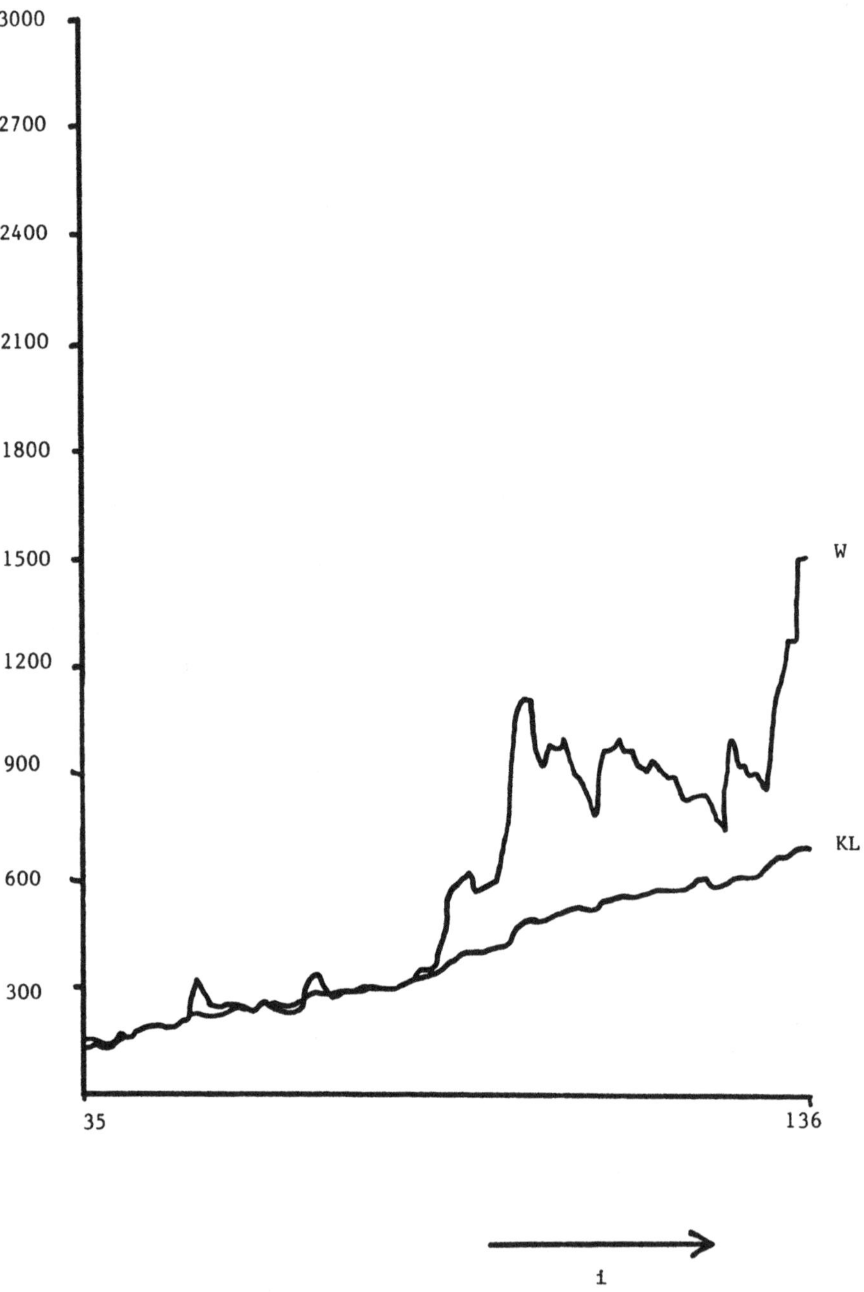

Figure 11 b. Median predictions from W and KL for data of Table 1. KL is very close to LV, so the latter is not plotted.

If we take the best six prediction systems (L, BL, LV, KL, LNHPP and W) and discount BL (because we cannot compute the medians), LV and KL (because they exhibit significant 'bias' as evidenced by the u-plot) we are left with L, LNHPP and W. The agreement between their median predictions is striking (see Figure 11).

What conclusions could a user draw from all this? We assume that he/she wishes to make predictions about the future and has available the data of Table 1 upon which our analysis is based. We also assume that he/she is prepared to trust for future predictions a prediction system whose past predictions have been shown to be close to actual observation (roughly this means that the future development of the software is similar to the past: there is no discontinuity between past and future). In that case *we would suggest that, for this data source, a user makes a future prediction using LNHPP.*

Notice that it is possible for the 'preferred prediction system' to change as the data vector gets larger. For example, based on the *whole* of the Table 1 data LV is preferable to JM; on the first 60 observations JM is better than LV (see Table 5). Thus, our advice to a user to use LNHPP is strictly only applicable for the next prediction. In principle this analysis should be repeated at each step. In practice this is sometimes not necessary: we have found that changes usually take place fairly slowly so that relatively infrequent checks on predictive quality are sufficient. However, these analyses are not computationally onerous when compared with the numerical optimisations needed for the ML estimation at each step.

Table 7 shows a similar analysis of the data of Table 2. The prequential likelihood suggests that L, BL, LV and KL perform best. A more detailed study shows an interesting trade-off between 'bias' and 'noise'. The KL and LV u-plot distances are significant and the plots are both below the line of unit slope, indicating that the predictions are *too pessimistic.* The noise statistics based on medians and rates show that L is more noisy than LV and KL. Since they give similar PL values we can conclude that L is *objectively* too noisy. We thus have an interpretation of the behaviour of the plots of Figure 12.

Model	$\Sigma\left\|\frac{m_i - m_{i-1}}{m_{i-1}}\right\|$ (rank)	$\Sigma\left\|\frac{r_i - r_{i-1}}{r_{i-1}}\right\|$ (rank)	Braun statistic (rank)	u-plot KS distance (sig. level)	y-plot KS distance (sig. level)	$-\log_e$ (PL), ranks
JM	4.23 (6)	3.81 (8)	1.11 (8)	.121 (NS)	.115 (NS)	466.22 (6)
BJM		3.27 (7)	1.04 (5)	.110 (NS)	.077 (NS)	466.64 (7)
L[(1)]	5.27 (7)	4.66 (9)	1.07 (7)	.123 (NS)	.091 (NS)	465.49 (2)
BL		2.82 (6)	.96 (1)	.138 (NS)	.068 (NS)	465.18 (1)
LV	2.33 (3)	2.18 (4)	.97 (3)	.167 (10%)	.051 (NS)	465.52 (3)
KL	2.33 (3)	2.17 (3)	.96 (1)	.170 (10%)	.051 (NS)	465.81 (4)
D	1.96 (2)	1.84 (2)	1.04 (5)	.209 (2%)	.052 (NS)	467.78 (9)
GO	1.06 (1)	1.03 (1)	1.21 (10)	.271 (1%)	.085 (NS)	473.87 (10)
LNHPP	3.05 (5)	2.76 (5)	.97 (3)	.169 (10%)	.082 (NS)	465.85 (5)
W	6.16 (8)	5.48 (10)	1.17 (9)	.100 (NS)	.111 (NS)	466.89 (8)

Table 7. Analysis of data of Table 2. Here [(3)] the ML routine did not terminate normally in a high proportion of cases for L and LV. The extreme closeness of LV and KL predictions (the latter always terminating normally) suggests that the LV values are close to optimal. This is not obvious for L, which may be able to give better results.

Figure 12. Median predictions from KL, LV and L for data of Table 2. KL and LV are identical to within the resolution of this table: less than 2% difference in medians.

A user is faced with an interesting choice between L and KL or LV, and the analysis helps him to exercise it intelligently. If he prefers to be conservative, and consistently so, he should use LV or KL. If he prefers to be closer to the true reliability *on average* (but with fluctuating errors) he should choose L. This is similar to the situation faced by the statistician, with a choice between two estimators having the same mse but different variances and bias (see 32).

There is evidence here that KL and LV are 'merely' biased. They are therefore good candidates for the adaptive ideas of the next section. There is some evidence that the predictions of KL and LV are almost identical for this data.

Table 8 shows the analysis of Table 3 data. Here BL seems to be giving the best results. It is, however, surprising that so many models do well on this data set. This may have something to do with the way the data was collected. The interfailure times refer to total operating time for a *population* of copies of the system. When a failure is observed, a fix (i.e. software fix or hardware design change) is introduced into all copies.

There are three successive very large observations near the end of the data set (each is larger than any previous observation in a data set with fairly slow reliability growth). Table 8 shows the PLR for *all* predictions, and for predictions which exclude these large observations. BL is best in both cases, but others change ranks dramatically. LV, for example does well on the smaller set of predictions, but poorly on the full set. This suggests that LV is assigning low probability density to the large observations (i.e. they lie in the tails of the predictive distributions). BJM, on the other hand, improves its rank equally dramatically by *including* the large observations.

If nothing else, this shows the importance of careful data collection. Here, we cannot know whether these large observations should be discounted or not.

Model	$\Sigma \left\lvert \frac{m_i - m_{i-1}}{m_{i-1}} \right\rvert$ (rank)	$\Sigma \left\lvert \frac{r_i - r_{i-1}}{r_{i-1}} \right\rvert$ (rank)	Braun statistic (rank)	u-plot KS distance (sig. level)	y-plot KS distance (sig. level)	$-\log_e$ (PL), rank[3]
JM	4.77 (7)	4.52 (9)	.941 (4)	.083 (NS)	.073 (NS)	711.0, 4 [637.2, 9]
BJM[2]		4.38 (7)	.954 (9)	0.84 (NS)	0.71 (NS)	710.9, 2 [637.5, 10]
L[1]	4.98 (8)	4.74 (10)	.943 (5)	.079 (NS)	.069 (NS)	711.1, 6 [637.0, 7]
BL		3.11 (4)	.935 (1)	.064 (NS)	.056 (NS)	710.4, 1 [635.9, 1]
LV	2.64 (2)	2.75 (2)	.949 (7)	.087 (NS)	.065 (NS)	711.8, 8 [636.3, 3]
KL	2.76 (3)	2.89 (3)	.953 (8)	.102 (NS)	.066 (NS)	712.6, 9 [636.8, 6]
D	1.98 (1)	1.92 (1)	.994 (10)	.117 (10%)	.074 (NS)	714.5, 10 [636.7, 5]
GO	4.58 (5)	4.30 (6)	.937 (2)	.075 (NS)	.075 (NS)	711.0, 4 [637.1, 8]
LNHPP	3.30 (4)	3.14 (5)	.939 (3)	.072 (NS)	.058 (NS)	710.9, 2 [636.1, 2]
W	4.67 (6)	4.43 (8)	.945 (6)	.064 (NS)	.057 (NS)	711.3, 7 [636.7, 4]

Table 8. Analysis of data of Table 3. Again, ML terminated abnormally for many L predictions[1]. BJM[2] calculations involved overflow on some predictions. Both these predictors may be able to give better results. The second set[3] of [$-\log_e$ (PL), rank] ignores <u>the last ten predictions</u>. There are three very large observations at the end (larger than any experienced before): the switches in rank indicate the effect these observations have upon predictive quality.

7. Adapting and Combining Predictions; Future Directions

In several data sets we have found that certain prediction systems seem to be 'only' biased: an example is LV for the data of Table 2. In cases like this, where the relationship between prediction and reality is approximately stationary, we can attempt to estimate this relationship and use the estimate to improve future predictions: a process of *recalibration*. Formally, we have

$$F_i(t) = G_i[\tilde{F}_i(t)]$$

where, at least approximately, G_i does not depend upon i. Since $\tilde{F}_i(t)$ is a function of data observed so far, the sequence G_i can be thought of as a random sequence of functions. The stationarity of the sequence is the key idea. If we are assured of this stationarity, we might reasonably attempt to find a 'best' estimate of the function, G_i^* say, to obtain an adapted predictor:

$$\tilde{F}_i^*(t) = G_i^*[\tilde{F}_i(t)] \tag{43}$$

Of course, this stationarity is a very strong condition and it would be unreasonable to expect it to hold exactly in practice. It may hold approximately. At the very least, it will be necessary to check that previous predictions have given a good y-plot. Better tests for approximate stationarity are needed and are being investigated.

Given that we believe there is approximate stationarity in the relationship between $\{\tilde{F}_i\}$ and $\{F_i\}$, there are various ways we might approach the problem of finding a 'best' G_i^*. One of the simplest is to use the u-plot, as shown in Figure 13.

The procedure is then as follows:

<u>Stage</u> <u>1</u>: Check that y-plot is good (based on $t_1, t_2, \ldots, t_{i-1}$).

<u>Stage</u> <u>2</u>: Find u-plot for predictions made <u>before</u> T_i, (i.e. based on subsets of $t_1, t_2, \ldots, t_{i-1}$).

<u>Stage</u> <u>3</u>: Use the prediction system to make a 'raw' prediction, $\tilde{F}_i(t)$.

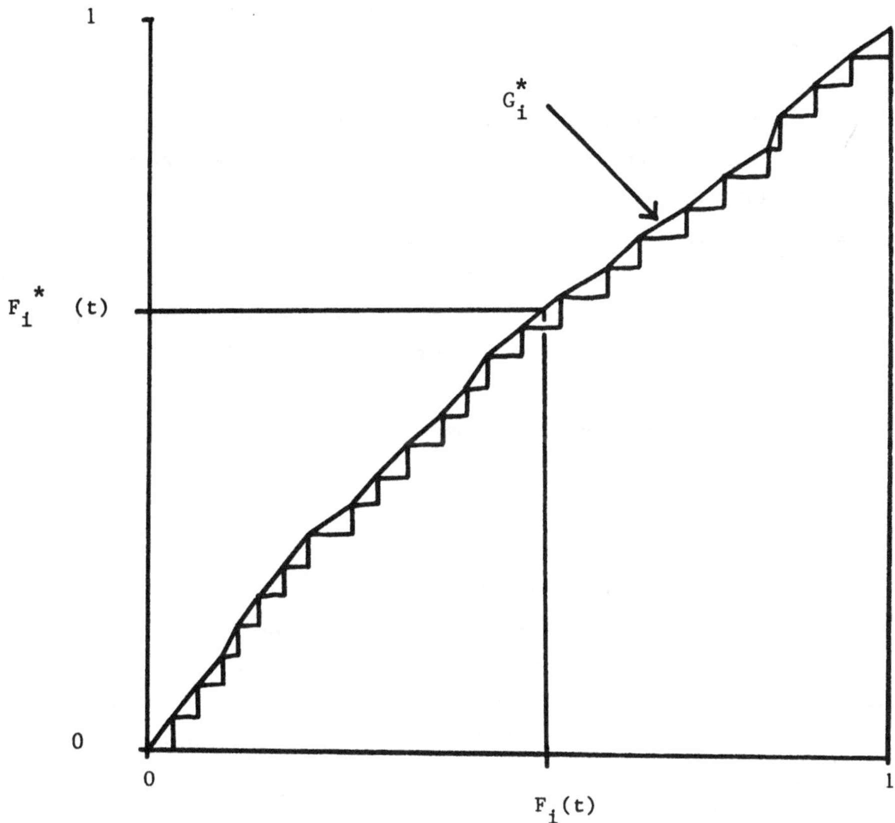

Figure 13. Method of calculating an adapted prediction. G_i^* is the 'joined-up' u-plot based on predictions made *prior to observing* the random variable T_i. $F_i(t)$ is the raw prediction we would make from the basic prediction system at this stage, $F_i^*(t)$ is the modification of this which 'learns' from the quality of previous predictions.
Notice that this figure is presented only for explanatory purposes. The computational form of the transformation of $F_i(t)$ into $F_i^*(t)$ is particularly simple.

Stage 4: Transform the raw prediction via the u-plot to obtain the adapted prediction, $\tilde{F}_i^*(t)$.

It is important to emphasise that this is a genuine prediction system again. At each stage we are obtaining $\tilde{F}_i^*(t)$ by using only information observed earlier. Thus we can use some of the devices of section 5 to examine the quality of the new prediction system.

Figure 14 shows the median of the adapted LV and adapted JM predictions for the data of Table 1. In each case it is obvious that the adapted version is better than the raw version: adapted LV is less pessimistic than the known pessimistic LV, adapted JM is less optimistic than known optimistic JM. It is clear that the noise of these median plots remains more or less unchanged. This is not surprising, since the $\{G_i^*\}$ sequence is 'slowly changing': successive G_i^* differ by the addition of one point to the u-plot.

The $\{\tilde{F}_i^*(t)\}$ sequence can be used to generate

$$u_i^* = \tilde{F}_i^*(t_i) \tag{44}$$

as before, since t_i (the realisation of T_i) is observed after $\tilde{F}_i^*$ is calculated. We can thus form u*-plots and y*-plots in the usual way. In this case the adapted LV has KS distances of 0.084 (u*-plots) and 0.078 (y*-plots) compared with 0.144 and 0.110. Adapted JM gives 0.104 and 0.108 compared with 0.190 and 0.120 for the raw JM. Not only are these an improvement in both cases, but now all four distances are *non-significant*.

It is surprising that in this case y*-plots are better than the y-plots. The procedure is only designed to improve 'bias'. This seems to be a common result of using the adaptive procedure. We have used it on several data sets with several different prediction procedures (Littlewood & Keiller (1984)). In no case is the u*-plot significantly worse than the u-plot, and it is usually better. In only one case out of the 30 examined was the y*-plot significantly worse than the y-plot, and again it was usually better.

Figure 15 shows the effect of applying the adaptive procedure to the LV analysis of the data of Table 2. As we detected in section 6, LV is slightly too pessimistic (significant u-plot KS distance, but only at 10% level). The adapted predictions are more optimistic,particularly near the end of the data.

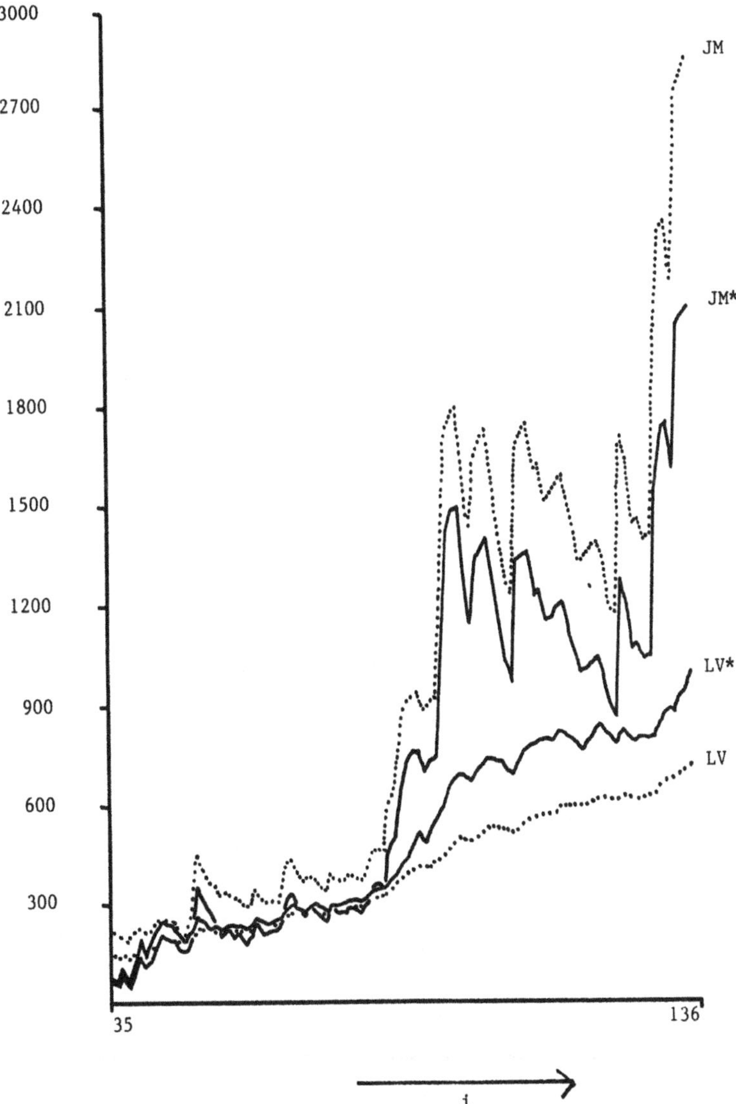

Figure 14. Median predictions for JM, LV, JM*, LV* on data of Table 1. Notice how the adaptive procedure involves a shift from the raw values without an obvious change in noisiness.
JM* is less optimistic than the known over-optimistic JM, LV* is less pessimistic than the known over-pessimistic LV.

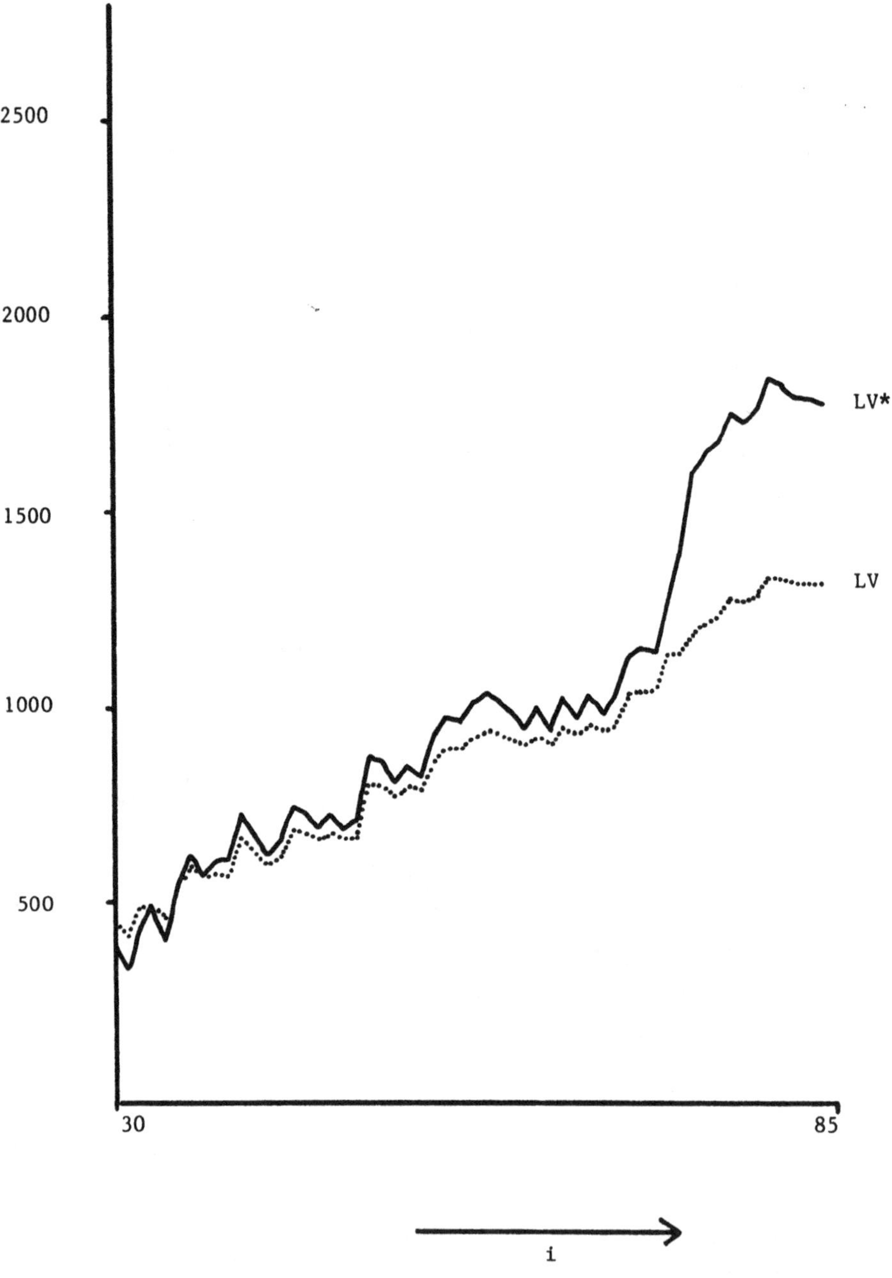

Figure 15. Median predictions for LV and LV* on data of Table 2.

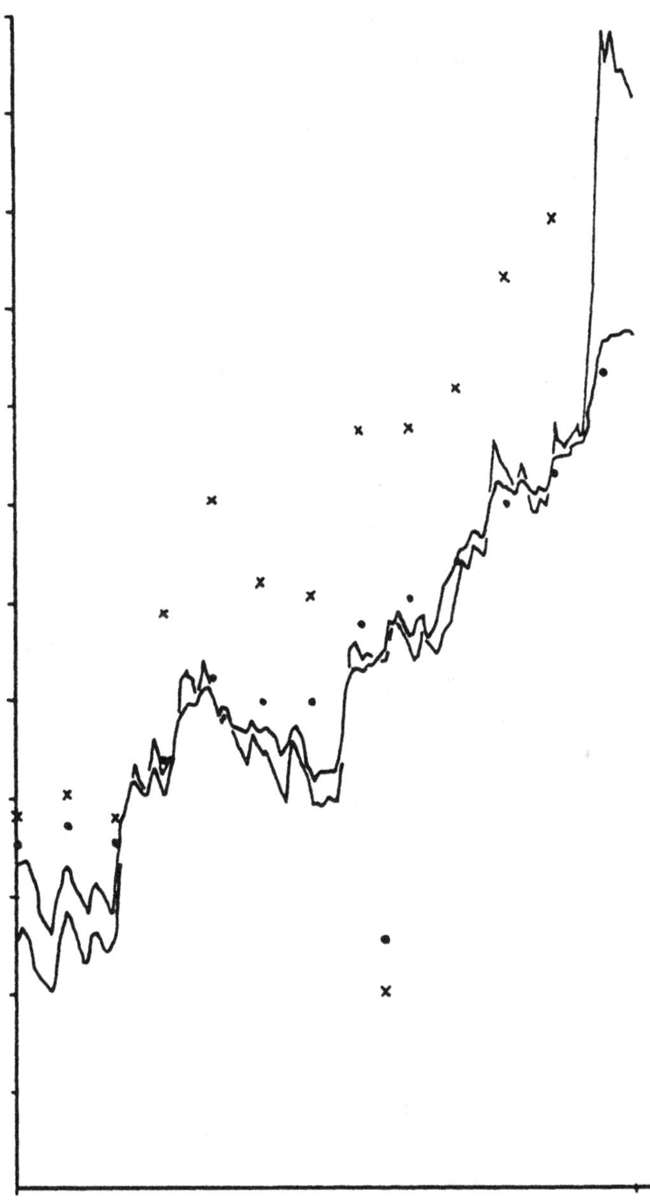

Figure 16. Median predictions for JM, LV, JM*, LV* on data of Table 3 (LV, JM only at 10–step intervals for clarity, JM* broken where it crosses the (unbroken) LV*).
Adaptive predictions are in very close agreement, except at very end. JM* noisier than LV*.

Figure 16 shows raw and adaptive LV and JM on the data of Table 3. Again, the adapted versions are in closer agreement, and differ from the raw predictions in the right way (as indicated by the u-plots).

We have here, then, a very simple procedure which can be used with any prediction system and will usually produce a better prediction system. Computational requirements are light, even for the step-by-step modifications of G_i^* used here.

There does not seem to be any price paid for the improvements which come from 'learning' from past behaviour. However, the timid user has available some of the devices of section 5 to check the quality of the adapted predictions.

Unfortunately, the prequential likelihood cannot be used here. Since the PL depends on predictive probability densities, it uses the slopes of the $\{G_i^*\}$ functions. Although we would expect the G_i^* functions to change slowly with i, the slopes (probability density function) for a particular G_i^* are highly variable. Thus a use of PLR to compare $\{\tilde{F}_i^*\}$ with $\{F_i\}$ would "unfairly" discriminate against $\tilde{F}_i^*$. This is seen for the data of Table 1, where -ln PL = 668.994 for raw JM predictions, but takes the greatly worse value of 680.743 for adapted predictions [Note: these figures are not comparable with those of Table 6, being based on 80 predictions rather than 100; this results from the first adapted prediction using the first 20 raw predictions to calculate the G_i^*].

This raises an interesting question: Most users would be happy to be able to make accurate probability predictions of future behaviour. PL, as have used it, requires accurate prediction of probability density. Since the joined-up u-plot estimate of G_i will give a discontinuous predictive density, PL is "correct" in rejecting this in favour of the biased but smooth raw predictor. However, there is a sense in which density predictions are of no practical interest and PL can be thought of as too strong a criterion for comparing adapted with raw predictions.

One way forward might be to construct a prequential likelihood based on prequential probabilities, but it is not obvious how to choose appropriate intervals on the time axis about which to make predictions.

Another approach is to assume that $G_i^* \in G$, some suitable parametric family, for example Beta distributions. This raises the interesting possibility of using PL as a means of estimating the transforming function, G_i^*, in the cases where this is differentiable. Since

$$\tilde{F}_i^*(t) = G_i^*[\tilde{F}_i(t)]$$

we have predictive densities

$$\tilde{f}_i^*(t) = g_i^*[\tilde{F}_i(t)]\ \tilde{f}_i(t) \tag{45}$$

where g_i^* is the derivative of G_i^* and $\tilde{f}_i(t)$ is the raw predictive density. Consider now the problem of modifying the raw predictor for T_{j+n+1} in the light of the performance of the prediction system in predicting T_{j+1}, T_{j+2}, ..., T_{j+n}. It seems sensible to choose our G_{j+n+1}^* to be the member of G which maximises the PL over the observed history from j + 1 to j + n :

$$\max_G \prod_{j+1}^{j+n} \tilde{f}_i(t_i)$$

$$= \max_G \prod_{j+1}^{j+n} \{g[\tilde{F}_i(t_i)]\}\ \tilde{f}_i(t_i)$$

$$= \prod_{j+1}^{j+n} \tilde{f}_i(t_i) \max_G \prod_{j+1}^{j+n} g(u_i) \tag{46}$$

That is, a procedure based on optimising prequential likelihood over previous predictions is equivalent to using maximum likelihood to estimate the (assumed common) distribution of the $\{U_i\}$ sequence based on u_{j+1}, ..., u_{j+n}. This very interesting observation throws light upon the adaptive procedure. It shows, for example, that our strong stationarity assumption is merely an assumption that $\{U_i\}$ have a common distribution: the problem then is simply one of estimating this distribution given some observations, u_i.

The drawback, of course, is that this loses the distribution-free flavour of the original adaptive approach. We have therefore chosen to simply smooth in order that the predictive densities are smooth. After trying various smoothing approaches, such as histo-splines, we found the most effective was a constrained least squares parametric spline for the joined-up u-plot estimator, G^*. Details of this can be found elsewhere (Chan *et al* (1985)). For the 80 predictions of this data of Table 1 this gives a value of –ln PL = 662.264 compared with 668.944 for the raw JM predictions: a considerable improvement. For the LV model on the same data the values of –ln PL are 663.348 for the raw model, 685.545 for the joined-up u-plot adaptor, and 659.846 for the spline-smoothed adaptor: once again the "naive" adaptor is unjustifiably discriminated against by the PL criterion, but the smoothing shows that in fact the recalibration is working very well. These results are typical of what we have found in analyses of several models on many data sets. The adaptive procedure using the joined-up u-plot appears to work well (in terms of improved u-plot for predictions after adapting) but the PL procedure reports unfavourably.

Brocklehurst (1987) conducted a simulation study to see whether the "naive" adaptive approach was producing predictive cdfs which were closer to the true cdf than the raw model. Here the distances between true, raw predictive and naive adapted cdfs could be measured (Kolmogorov distances were generally used). More than 90% of cases showed the naive adapter being superior to the raw predictor in those cases where the original u-plot was significant (i.e. there was evidence of bias and so *a priori* justification for using the adaptive procedure). Surprisingly, the efficacy of the adaptive approach did not seem to be critically dependent upon the non-significance of the y-plots. The result of this simulation analysis seems to be that the simple adaptive procedure usually works well, but that any user who needed confirmation that it was working well in a particular context would need to smooth in order to conduct a PL analysis.

There are other ways in which we could learn from past predictive behaviour on a particular data source in order to improve future predictions. One approach is to combine predictions arising from different (perhaps themselves adaptive) prediction systems. Given prediction systems A and B we might consider a predictor.

$$\tilde{F}^{(A+B)}_{j+n+1}(t) = w_A \tilde{F}^{A}_{j+n+1}(t) + w_B \tilde{F}^{B}_{j+n+1}(t) \qquad (47)$$

where w_A (= 1 – w_B) would be chosen in some optimal way in the light of experience of previous predictions emanating from A and B.

We have tried two ways of combining predictors. The first method is motivated by how we would behave if we were in a pure Bayesian context. If $\{\tilde{F}_i^A\}$ and $\{\tilde{F}_i^B\}$ were Bayesian predictive distributions, we would form the 'meta' Bayesian predictor, (46), by letting w_A, w_B be the posterior probabilities of models A and B.

Of course, it is rare to have two Bayesian prediction systems. We shall therefore appeal to the argument of section 5.4 which showed that the PLR can be interpreted as the approximate posterior odds of A against B (see (42)). Thus at stage j+n+1 in (46) we shall use the value w_A obtained from

$$PLR = \frac{w_A}{1 - w_A}$$

where PLR is based on predictions j + 1 through j + n. We shall call this meta procedure $(A + B)_1$. It was suggested by Ian Goudie (1984).

An alternative approach is to choose that value of w_A to use in (47) which would have maximised the PL for $\{\tilde{F}_i^{(A+B)}, \ i = j+1, \ldots, j+n\}$. We shall refer to this prediction system as $(A + B)_2$.

We emphasise again that these procedures are proper prediction systems. They rely only upon data collected in the past, which is used dynamically to change the weights w_A, w_B at each stage. Once again, a user does not need to believe *a priori* that $(A + B)_1$ or $(A + B)_2$ are good: their predictive performance can be monitored, using the techniques of section 5, just like any other prediction system.

Figure 17 shows median plots for $(LV+JM)_1$ and $(LV+JM)_2$ predictions of the data of Table 1. There is a tendency for $(LV+JM)_1$ to give very small weights to JM throughout the range of predictions. This is reflected in the median plots, where $(LV+JM)_1$ and LV are almost identical. As for $(LV+JM)_2$, this gives weight 0 exactly to JM for predictions 40 through 84. This is again seen in the median plot.

The PL for the four prediction systems suggests that $(LV+JM)_2$ is significantly better than the other three. The u- and y-plot KS distances are given in Table 9. It can be seen that $(LV+JM)_2$ is best on each.

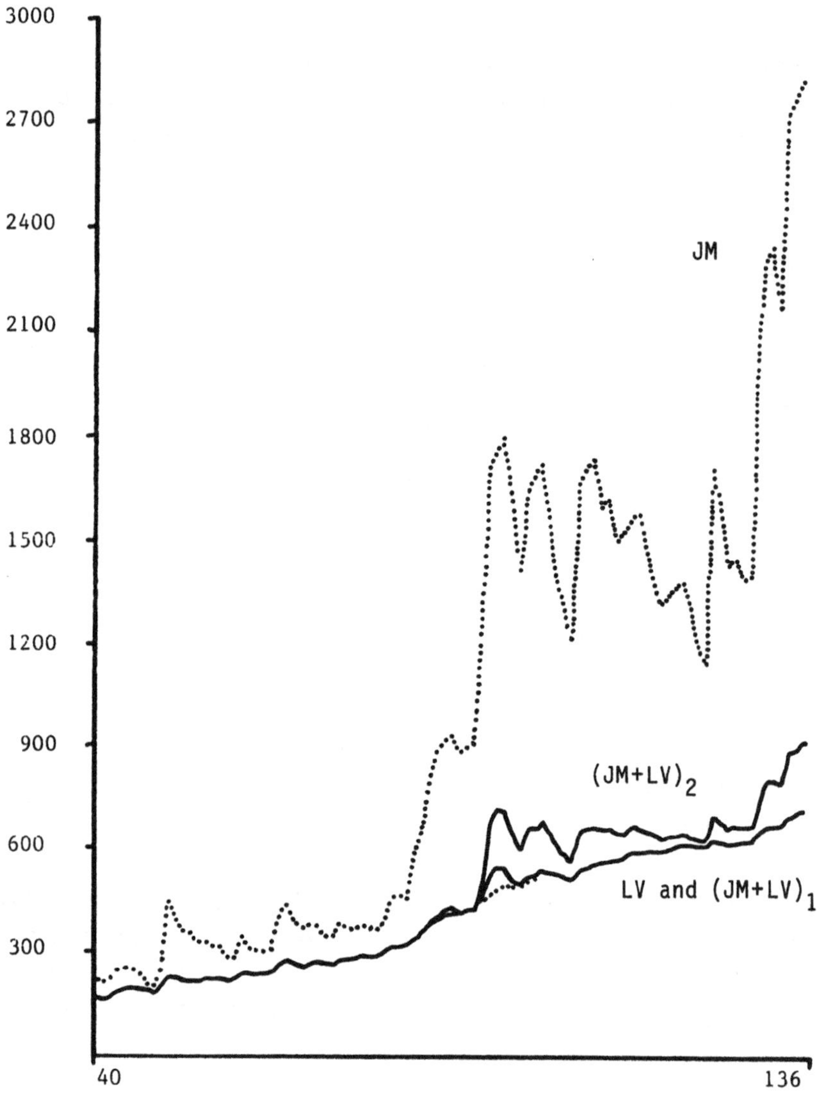

Figure 17. Median predictions from JM, LV, $(JM+LV)_1$ and $(JM+LV)_2$ for the data of Table 1. Both combined models are very close to, or identical to, LV for the early predictions.

Model	KS distances	
	u–plot	y–plot
JM	.189	.126
LV	.145	.107
(LV+JM)	.145	.105
(LV+JM)	.134	.076

Table 9. KS distances for u– and y–plots of JM, LV and combined predictions, data of Table 1. Notice that these refer to a slightly different range of i, so the results do not correspond exactly to those of Table 6.

An analysis of this data using $(LV+JM)_2$ is interesting. Here $w_{JM} = 0$ for all predictions, i.e. the 'combined' prediction totally rejects JM. This is plausible since both L and JM are optimistic, but L is less so; also L is less noisy than JM. Thus, for both 'bias' and 'noise' criteria, L is better than JM, and there is no benefit to be gained by incorporating the JM predictions. This contrasts with LV and JM, where we know that one is pessimistic, the other optimistic (particularly for later predictions). There is therefore benefit in using a weighted average, since the truth is known to lie between the two raw predictions.

Both the adaptive and combined predictions described here seem attractive candidates for further study. It would be easy to combine more than two predictions, using either of our two criteria. In fact we can think of generalisations to super meta prediction systems. For example, we could adapt all models first and then combine the adapted models.

In this paper we have concentrated upon prediction of the next time to failure. Clearly, some of the techniques described here can be used for longer term prediction. We could, for example, predict T_{i+20}, based on data t_1, t_2, ..., t_{i-1} and analyse these predictions using the ideas of section 5.

It is fairly obvious that a prediction system which performs well for short-term prediction cannot be guaranteed to perform well for longer-term prediction. In Figure 4, for example, JM does dramatically worse for 20-step ahead predictions than it does for 1-step ahead. This suggests a 'horses for courses' approach: a user should be sure that a prediction system is performing well for the *type of prediction* of interest.

Clearly, there are certain types of prediction which are not obviously amenable to the analysis described in section 5. For example, it is commonly desired to predict the time to achieve a specified target reliability (perhaps the reliability which must be achieved before the product can be shipped). It is not clear how predictions of this (random variable) time can be analysed *directly*. They could be examined *indirectly* via an analysis of predictions of T_i, T_{i+1}, ... based on t_1, ..., t_{i-1}, but this may not be adequate. Problems of this kind require further study.

8. Summary and Conclusions

Our theme in this paper has been that predictive quality is the only consideration for a user of software reliability models. We have shown that a good 'model' is not sufficient to produce good predictions, and that there is no universal 'best buy' among the competing methods. Prediction systems perform with varying adequacy on different data sources. Users, therefore, need to be able to select, from a plethora of prediction systems, one (or more) which gives trustworthy predictions for the software being studied. Our aim in this work has been to provide some tools to assist this selection. We do not claim that these are complete; indeed, we hope that other workers will be stimulated to find different techniques. In the meantime, we believe that the work reported here can be used by anyone wishing to predict the reliability of software.

If workers in the community can agree on suitable criteria for quality of prediction, there arise possibilities for new approaches via adaptive and learning techniques. We believe, from our experience with rudimentary adaptive modelling, that these ideas show great promise. Not merely do they allow current models to be improved in very general ways, but they also point the way forward to new methods with weak underlying assumptions.

Acknowledgements

This paper summarises recent work which has been conducted by several people in the Centre for Software Reliability at City University, and elsewhere. It is a pleasure to acknowledge the contribution of past and present members of the staff of CSR, particularly P.Y. Chan, Abdallah Abdel-Ghaly, and Sarah Brocklehurst, as well as John Snell of the City University Department of Computer Science. The author has benefitted greatly from discussions with many people over the past few years, but would like to single out Tom Anderson, Phil Dawid, Mike Dyer, Ian Goudie, Barbara Kitchenham, Peter Mellor, Earle Migneault, Doug Miller, John Musa and Ariela Sofer.

This work has been supported over the years by grants from ICL Ltd, National Aeronautics and Space Administration, the Science and Engineering Research Council and the Alvey Directorate.

References

ABDEL GHALY, A.A. (1986), Ph D Thesis, City University, London.

ADAMS, E.N. (1984), Optimizing preventive service of software products, *IBM Journal of Research and Development*, **28** (No 1).

AITCHISON, J. & DUNSMORE, I.R. (1975), *Statistical Prediction Analysis*, Cambridge University Press, Cambridge.

AKAIKE, H. (1982), Prediction and Entropy, *MRC Technical Summary Report*, Mathematics Research Center, University of Wisconsin–Madison.

AMMAN, P.E. & KNIGHT, J.C. (1987), Data diversity: an approach to software fault tolerance, *Digest FTCS*-17 (17th International Symposium on Fault-Tolerant Computing), 122–126.

ASCHER, H. & FEINGOLD, H. (1984), *Repairable Systems Reliability*, Marcel Dekker, New York.

BRAUN, H. & PAINE, J.M. (1977), A comparative study of models for reliability growth, *Technical Report No* 126, Series 2, Department of Statistics, Princeton University.

BROCKLEHURST, S. (1987), On the effectiveness of adaptive software reliability modelling, *CSR Technical Report*, City University, London.

CHAN, P.Y., LITTLEWOOD, B. & SNELL, J. (1985), Parametric spline approach to adaptive reliability modelling, *CSR Technical Report*, City University, London.

COX, D.R. & LEWIS, P.A.W. (1966), *The Statistical Analysis of Series of Events*, Methuen, London.

CROW, L.H. (1977), Confidence interval procedures for reliability growth analysis, *Technical Report No* 197, US Army Material Systems Analysis Activity, Aberdeen, Md.

DALE, C.J. (1982), Software Reliability Evaluation Methods, *British Aerospace Dynamics Group*, **ST-26750.**

DAWID, A.P. (1982), The well-calibrated Bayesian, (with discussion),*Journal of the American Statistical Association*, **77**, 605–613.

DAWID, A.P. (1984 a), Calibration-based empirical probability, *Research Report* 36, Department of Statistical Science, University College, London.

DAWID, A.P. (1984 b), Statistical theory: the prequential approach, *Journal of the Royal Statistical Society, A* , **147**, 278–292.

DAWID, A.P. (1989), Probability Forecasting. In: S. Kotz, N. L. Johnson and C. B. Read (Eds), *Encyclopedia of Statistical Sciences*, Vol 6, Wiley, New York.

DUANE, J.T. (1964), Learning curve approach to reliability monitoring, *IEEE Transactions on Aerospace*, **2**, 563–566.

FORMAN, E.H. & SINGPURWALLA, N.D. (1977), An empirical stopping rule for debugging and testing computer software, *Journal of the American Statistical Association*, **72**, 750–757.

GOEL, A.L. & OKUMOTO, K. (1979), Time-dependent error-detection rate model for software reliability and other performance measures, *IEEE Transactions on Reliability*, **28**, 206–211.

GOUDIE, I. (1984), private communication.

JELINSKI, Z. & MORANDA, P.B. (1972), Software reliability research. In:W FREIBERGER (Ed), *Statistical Computer Performance Evaluation*, Academic Press, New York, 465–484.

JOE, H. & REID, N. (1985), Estimating the number of faults in a system, *Journal of the American Statistical Association*, **80**, 222–226.

KEILLER, P.A., LITTLEWOOD, B., MILLER, D.R. & SOFER, A. (1983 a), On the quality of software reliability predictions, *Proc. NATO ASI on Electronic Systems Effectiveness and Life Cycle Costing* (Norwich, UK, 1982), Springer, Berlin, 441–460.

KEILLER, P.A., LITTLEWOOD, B., MILLER, D.R. & SOFER, A. (1983 b), Comparison of software reliability predictions, *Digest FTCS* 13 (13th International Symposium on Fault-Tolerant Computing), 128–134.

KENDALL, M.G. & STUART, A. (1961), *The Advanced Theory of Statistics*, Griffin, London.

LANGBERG, N. & SINGPURWALLA, N.D. (1981), A unification of some software reliability models via the Bayesian approach, *Technical Report, TM*–66571, The George Washington University, Washington DC.

LAPRIE, J.C. (1984), Dependability evaluation of software systems in operation, *IEEE Transactions on Software Engineering*, **10.**

LITTLEWOOD, B. (1979), How to measure software reliability and how not to, *IEEE Transactions on Reliability*, **28**, 103–110.

LITTLEWOOD, B. (1981), Stochastic reliability growth: a model for fault–removal in computer programs and hardware designs, *IEEE Transactions on Reliability*, **30**, 313–320.

LITTLEWOOD, B. & KEILLER, P.A. (1984), Adaptive software reliability modelling, *Digest FTCS*–14 (14th International Conference on Fault–Tolerant Computing), 108–113.

LITTLEWOOD, B. & SOFER, A. (1985), A Bayesian modification to the Jelinski–Moranda software reliability growth model, *CSR Technical Report.*

LITTLEWOOD, B. & VERRALL, J.L. (1973), A Bayesian reliability growth model for computer software, *Applied Statistics*), **22**, 332–346.

LITTLEWOOD, B. & VERRALL, J.L. (1981), On the likelihood function of a debugging model for computer software reliability, *IEEE Transactions on Reliability*, **30**, 145–148.

MILLER, D.R. (1983), private communication.

MILLER, D.R. (1986), Exponential order statistic models of software reliability growth, *IEEE Transactions on Software Engineering*, **12,** 12–24.

MUSA, J.D. (1975), A theory of software reliability and its application, *IEEE Transactions on Software Engineering*, 1, 312–327.

MUSA, J.D. (1979), Software reliability data, report available from Data and Analysis Center for Software, Rome Air Development Center, Rome, NY.

NAGEL, P.M. & SKRIVAN, J.A. (1981), Software reliability: repetitive run experimentation and modelling, **BCS–40399** (Dec.), Boeing Computer Services Company, Seattle, Washington.

ROSENBLATT, M. (1952), Remarks on a multivariate transformation, *Annals of Mathematical Statistics*, **23**, 470–472.

SHOOMAN, M. (1973), Operational testing and software reliability during program development, in *Record.* 1973 *IEEE Symp. Computer Software Reliability* (New York, 1973, April 30 – May 2), 51–57.

CENTRE FOR SOFTWARE RELIABILITY

CITY UNIVERSITY

NORTHAMPTON SQUARE

LONDON EC1V 0HB

UK

REFERENCES

ABDEL GHALY, A.A. (1986), Ph D Thesis, City University, London.

ADAMS, E.N. (1984), Optimizing preventive service of software products, *IBM Journal of Research and Development*, **28** (No 1).

AITCHISON, J. & DUNSMORE, I.R. (1975). *Statistical Prediction Analysis*. Cambridge University Press.

AKAIKE, H. (1982), Prediction and Entropy, *MRC Technical Summary Report*, Mathematics Research Center, University of Wisconsin–Madison.

AMMAN, P.E. & KNIGHT, J.C. (1987), Data diversity: an approach to software fault tolerance, *Digest FTCS*-17 (17th International Symposium on Fault–Tolerant Computing), 122–126.

ASCHER, H. & FEINGOLD, H. (1984), *Repairable Systems Reliability*, Marcel Dekker, New York.

BALABAN, H.S. (1978), Reliability growth models, *Journal of Environmental Science*, 11–18.

BEACH, L.R., CHRISTENSEN-SZALANSKI, J. & BARNES, V. (1987). Assessing human judgement: has it been done, can it be done, should it be done? In: G. WRIGHT & P. AYTON (Eds), *Judgemental Forecasting*, John Wiley, Chichester, 49 – 62.

BRAU N , H. & PAINE, J.M. (1977), A comparative study of models for reliability growth, *Technical Report No* 126, Series 2, Department of Statistics, Princeton University.

BROCKLEHURST, S. (1987), On the effectiveness of adaptive software reliability modelling, *CSR Technical Report*, City University, London.

BUNDAY, B.D., AL–AYOUBI, I.D. & NEWBY, M.J. (1990), Likelihood and Bayesian Estimation Methods for Poisson Process Models in Software Reliability, *International Journal of Quality and Reliability Management*, **7** (5), 9–18.

CHAN, P.Y., LITTLEWOOD, B. & SNELL, J. (1985), Parametric spline approach to adaptive reliability modelling, *CSR Technical Report*, City University, London.

COOKE, R.M. (1987). A theory of weights for combining expert opinion. Department of Mathematics, Delft University of Technology.

COOKE, R.M. (1991). *Experts in Uncertainty: Expert Opinion and Subjective Probability in Science*. Oxford University Press.

COOKE, R.M., MENDEL, M. & THEYS, W. (1988). Calibration and information in expert resolution: a classical approach, *Automatica*, **24**, 87 – 94.

COOKE, R.M., STOBBELAAR, M. & VAN STEEN, J. (1988). Expert Opinion in Safety Studies: Case Report 4 – DSM Case. Department of Mathematics, Delft University of Technology.

COX, D.R. & LEWIS, P.A.W. (1966), *The Statistical Analysis of Series of Events*, Methuen, London.

COX, D.R. & LEWIS, P.A.W. (1978), *The Statistical Analysis of Series of Events*, Chapman and Hall, London.

CROW, L.H. (1974), Reliability analysis for complex repairable systems. In: F. PROSCHAN & R.J. SERFLING (Eds), *Reliability and Biometry*, 379–410, SIAM Philadelphia.

CROW, L.H. (1977), Confidence interval procedures for reliability growth analysis, *Technical Report No* 197, US Army Material Systems Analysis Activity, Aberdeen, Md.

DALE, C.J. (1982), Software Reliability Evaluation Methods, *British Aerospace Dynamics Group*, **ST-26750**.

DAWID, A.P. (1982), The well–calibrated Bayesian, (with discussion),*Journal of the American Statistical Association*, **77**, 605–613.

DAWID, A.P. (1984), Calibration–based empirical probability, *Research Report* 36, Department of Statistical Science, University College, London.

DAWID, A.P. (1984), Statistical theory: the prequential approach, *Journal of the Royal Statistical Society, A* , **147**, 278–292.

DAWID, A.P. (1989), Probability Forecasting. In: S. KOTZ, N. L. JOHNSON & C. B. READ (Eds), *Encyclopedia of Statistical Sciences*, Vol 6, Wiley–Interscience, New York.

DEGROOT, M.H. (1970). *Optimal Statistical Decisions*. McGraw–Hill, New York.

DUANE, J.T. (1964), Learning curve approach to reliability monitoring, *IEEE Transactions on Aerospace*, **2**, 563–566.

European Safety and Reliability Research and Development Association (1990). *Expert Judgement in Risk and Reliability Analysis: Experience and Perspective.* ESSRDA Report No. 2.

FORMAN, E.H. & SINGPURWALLA, N.D. (1977), An empirical stopping rule for debugging and testing computer software, *Journal of the American Statistical Association*, **72**, 750–757.

FRENCH, S. (1985). Group consensus probability distributions: a critical survey. In: J.M. BERNARDO, M.H. DEGROOT, D.V. LINDLEY & A.F.M. SMITH (Eds), *Bayesian Statistics* 2, North Holland, Amsterdam, 183 – 201.

FRENCH, S. (1987). Conflict of belief: when advisers disagree. In: P.G. BENNETT (Ed), *Analysing Conflict and its Resolution*, Oxford University Press.

FRENCH, S. & WIPER, M. (1990). Bayesian Updating by remodelling an Expert's Quantiles. Report 90.7, School of Computer Studies, University of Leeds.

GELFAND, A.E. & SMITH, A.F.M. (1990), Sampling based approaches to calculating marginal densities, *Journal of the American Statistical Association*, **85**, 398–409.

GENEST, C. & WAGNER, C. (1984). Further evidence against independence preservation in expert judgement synthesis.

GOEL, A.L. & OKUMOTO, K. (1979), Time-dependent error-detection rate model for software reliability and other performance measures, *IEEE Transactions on Reliability*, **28**, 206–211.

GOUDIE, I. (1984), private communication.

JELINSKI, Z. & MORANDA, P.B. (1972), Software reliability research. In: W. FREIBERGER (Ed), *Statistical Computer Performance Evaluation* , Academic Press, New York, 465–484.

JOE, H. & REID, N. (1985), Estimating the number of faults in a system, *Journal of the American Statistical Association*, **80**, 222–226.

KAHNEMAN, D., SLOVIC, P. & TVERSKY, A. (Eds) (1982). *Judgement under Uncertainty: Heuristics and Biases.* Cambridge University Press.

KEILLER, P.A., LITTLEWOOD, B., MILLER, D.R. & SOFER, A. (1983), On the quality of software reliability predictions, *Proc. NATO ASI on Electronic Systems Effectiveness and Life Cycle Costing* (Norwich, UK, 1982), Springer, Berlin, 441–460.

KEILLER, P.A., LITTLEWOOD, B., MILLER, D.R. & SOFER, A. (1983), Comparison of software reliability predictions, *Digest FTCS* 13 (13th International Symposium on Fault-Tolerant Computing), 128–134.

KENDALL, M.G. & STUART, A. (1961), *The Advanced Theory of Statistics*, Griffin, London.

LANGBERG, N. & SINGPURWALLA, N.D. (1981), A unification of some software reliability models via the Bayesian approach, *Technical Report*, *TM*-66571, The George Washington University, Washington DC.

LAPRIE, J.C. (1984), Dependability evaluation of software systems in operation, *IEEE Transactions on Software Engineering*, **10.**

LICHTENSTEIN, S. & FISCHOFF, B. (1980). Training for calibration, *Organisational Behaviour and Human Performance*, **28**, 149 - 171.

LICHTENSTEIN, S., FISCHOFF, B. & PHILLIPS, L.D. (1982). Calibration of probabilities: the state of the art until 1980. In: KAHNEMAN *et al*, 306 – 334.

LINDLEY, D.V. (1985), *Making Decisions (Second Edition)*, Wiley, New York.

LITTLEWOOD, B. (1979), How to measure software reliability and how not to, *IEEE Transactions on Reliability*, **28**, 103–110.

LITTLEWOOD, B. (1981), Stochastic reliability growth: a model for fault-removal in computer programs and hardware designs, *IEEE Transactions on Reliability*, **30**, 313–320.

LITTLEWOOD, B. & KEILLER, P.A. (1984), Adaptive software reliability modelling, *Digest FTCS*-14 (14th International Conference on Fault-Tolerant Computing), 108–113.

LITTLEWOOD, B. & SOFER, A. (1985), A Bayesian modification to the Jelinski-Moranda software reliability growth model, *CSR Technical Report*.

LITTLEWOOD, B. & VERRALL, J.L. (1973), A Bayesian reliability growth model for computer software, *Applied Statistics*, **22**, 332–346.

LITTLEWOOD, B. & VERRALL, J.L. (1981), On the likelihood function of a debugging model for computer software reliability, *IEEE Transactions on Reliability*, **30**, 145–148.

MANN, N.R., SCHAFER, R.E. & SINGPURWALLA, N.D. (1974), *Methods for the Statistical Analysis of Reliability and Life Data*, John Wiley, New York.

MARTZ, H.J. & WALLER, R.A. (1982), *Bayesian Reliability Analysis*, John Wiley, New York.

MENDEL, M.B. & SHERIDAN, T. (1987). Optimal estimation using human experts. Man-machine Systems Lab., Department of Mechanical Engineering, MIT.

MERKHOFER, M.W. (1987). Quantifying judgemental uncertainty: methodology experiences and insights, *IEEE Transactions on Systems, Man and Cybernetics*, **17**, 741 – 752.

MILLER, D.R. (1983), private communication.

MILLER, D.R. (1986), Exponential order statistic models of software reliability growth, *IEEE Transactions on Software Engineering*, **12,** 12–24.

MORRIS, P.A. (1974). Decision analysis: expert use, *Management Science*, **20,** 1233 - 1241.

MUSA, J.D. (1975), A theory of software reliability and its application, *IEEE Transactions on Software Engineering*, **1,** 312–327.

MUSA, J.D. (1979), Software reliability data, report available from Data and Analysis Center for Software, Rome Air Development Center, Rome, NY.

NAGEL, P.M. & SKRIVAN, J.A. (1981), Software reliability: repetitive run experimentation and modelling, **BCS-40399** (Dec.), Boeing Computer Services Company, Seattle, Washington.

NAYLOR, J.C. & SMITH, A.F.M. (1982), Applications of a method for the efficient computation of posterior distributions, *Applied Statistics*, **31,** 214–225.

RAIFFA, H. (1968). *Decision Analysis*. Addison Wesley, Reading, Mass.

ROSENBLATT, M. (1952), Remarks on a multivariate transformation, *Annals of Mathematical Statistics*, **23**, 470–472.

RUBIN, D.B. (1988). Using the SIR algorithm to simulate posterior distributions. In: J.M. BERNARDO, *et al* (Eds), *Bayesian Statistics* 3, Oxford University Press.

SHOOMAN, M. (1973), Operational testing and software reliability during program development, in *Record.* 1973 *IEEE Symp. Computer Software Reliability* (New York, 1973, April 30 – May 2), 51–57.

SMITH, A.F.M., SKENE, J.E.H. & NAYLOR, J.C. (1987), Progress with numerical and graphical methods for practical Bayesian statistics, *Statistician*, **36,** 75–82.

SMITH, A.F.M., SKENE, A.M., SHAW, J.E.H. & NAYLOR, J.C. (1987), Progress with numerical and graphical methods for practical Bayesian statistics, *The Statistician*, 36, 75–83.

TNO (1982), Veiligheidsstudie betreffende het transport per ondergrondse pijpleiding van aardgas en LPG in Nederland (in Dutch).

WIPER, M.P. (1990). *Calibration and Use of Expert Probability Judgements.* PhD Thesis, School of Computer Studies, University of Leeds.

AUTHOR INDEX

SUBJECT INDEX

TOPICS IN SAFETY, RELIABILITY AND QUALITY

1. P. Sander and R. Badoux (eds.): *Bayesian Methods in Reliability*. 1991
ISBN 0-7923-1414-X

KLUWER ACADEMIC PUBLISHERS – DORDRECHT / BOSTON / LONDON

Zeitfracht Medien GmbH
Ferdinand-Jühlke-Straße 7
99095 Erfurt, Deutschland
produktsicherheit@kolibri360.de